MOLECULAR RADIOBIOLOGY

N. B. STRAZHEVSKAYA. Editor

MOLECULAR RADIOBIOLOGY

Translated from Russian by A.Mercado
Translation edited by P.Harry

A HALSTED PRESS BOOK

JOHN WILEY & SONS
New York · Toronto

ISRAEL PROGRAM FOR SCIENTIFIC TRANSLATIONS
Jerusalem · London

© 1975 Keter Publishing House Jerusalem Ltd.

Sole distributors for the Western Hemisphere
HALSTED PRESS, a division of
JOHN WILEY & SONS, INC., NEW YORK

Library of Congress Cataloging in Publication Data
Main entry under title:

Molecular radiobiology.

"A Halsted Press book."
Includes bibliographies.
1. Radiobiology. 2. Molecular biology.
I. Strazhevskaíà, N. B., ed.
QH652.M6413   574.1'915   75-1262
ISBN 0-470-83326-2

Distributors for the U.K., Europe, Africa and the Middle East
JOHN WILEY & SONS, LTD., CHICHESTER

Distributors for Japan, Southeast Asia, and India
TOPPAN COMPANY, LTD., TOKYO AND SINGAPORE

Distributed in the rest of the world by
KETER PUBLISHING HOUSE JERUSALEM LTD.
ISBN 0 7065 1432 7
IPST cat. no. 22114 3

This book is a translation from Russian of
MOLEKULARNAYA RADIOBIOLOGIYA
(SOVREMENNYE PROBLEMY RADIOBIOLOGII, TOM III)
Atomizdat
Moscow, 1972

Printed and bound by Keterpress Enterprises, Jerusalem
Printed in Israel

CONTENTS

INTRODUCTION

One of the major achievements of modern biology is the deep insight gained into the nature of vital phenomena at the molecular level. Molecular biology — a science concerned with the border region between organic macromolecular chemistry, biochemistry, biophysics, and genetics — has revealed the structure and functions of the macromolecules participating in the fundamental processes of life. It has shed light on the physical and chemical nature of such biological phenomena as heredity, development and the differentiation of living organisms, and has explained the mechanisms of protein and enzyme synthesis, as well as the formation of the fundamental structures of the cell. While closely linked with physics and chemistry, modern biology is increasingly using mathematical methods to analyze vital phenomena and utilizes the concepts of cybernetics in order to understand the self-regulating processes underlying the existence of any living system.

These front line topics of biology are reflected in many specialized branches of this science, including radiobiology.

The history of the recent science of radiobiology can be divided into three phases. The first phase, which began with the discovery of ionizing radiations, comprised the accumulation of data concerning the effects of irradiation on various biological objects. This phase can be characterized as the embryonic period in the development of radiobiology — a time when radiobiology was not yet a true science but its rudiments were taking shape within the framework of other disciplines.

Radiobiology became a science in its own right in the 1940s, largely as a result of the rapid advances in nuclear physics, which gave rise to the second phase. The basic quantitative relationships between the energy absorbed and the final radiobiological events were established at that time. The conception of the observed effects of radiation at the molecular and cellular levels as statistical phenomena was expressed in the target theory, a mathematical model enabling the size of certain homogeneous molecules and simple viruses to be calculated during their irradiation.

Using the concepts of the target theory, radiobiologists strove to elucidate the primary processes underlying the observed phenomena by analyzing dose curves as a function of the physical parameters of the radiation. Hence the concepts of single and multiple hit mechanisms and the establishment of single and multiple target models. With the progress of knowledge about the structures and processes involved in such a complex system as the cell, however, the originally clear-cut principles of the target theory are becoming increasingly abstract and indefinite. The idea of the target as a particular molecule or structure is evolving into an abstract concept of a sensitive area as a rule unrelated to specific structures. The unequivocal principle of the "all or none" effect of radiation — a natural consequence of the idea of hitting a particular target — is being gradually replaced by the concept of "latent lesions," the manifestation of which depends on various factors acting after the irradiation of the object.

To cross the line between passive observation and actual control of radiobiological events, scientists had to elucidate the true nature of the physicochemical, structural and metabolic processes occurring in the irradiated cell. The abstract concepts of the target theory, applicable to only a restricted number of events, had to give way before the new insights provided by recent progress in molecular biology, biochemistry and cellular biophysics.

Radiobiology thus entered the third phase of its development. Well equipped with modern biophysical and biochemical procedures for the analysis of the irradiated cell, it was now fully capable of formulating and solving the problems concerning the nature of the molecular processes in the complex, constantly metabolizing cell. By formulating a structural and metabolic theory of the effect of ionizing radiations on living objects, molecular radiobiology has gained a leading place in modern radiobiology.

What are the most crucial and urgent problems facing molecular radiobiology today? We shall not attempt to answer this difficult question in full but rather to draw attention to some of its aspects.

The main problem of radiobiology undoubtedly concerns the effect of radiation on the giant molecules of deoxyribonucleic acid (DNA), the molecules holding all the information required for the existence and development of the cell and of the organism as a whole (Chapters 1, 4, 5). Molecular radiobiology has by now accumulated a vast amount of experimental data concerning the nature of radiation-induced lesions of DNA, the radiation yield of these lesions and the effect of various factors on the manifestation of these radiation lesions during and after irradiation. These data, obtained by radiochemical experiments on isolated DNA molecules, are tested by molecular radiobiology in the living, metabolizing cell. The radiochemical data are thus modified and corrected according to the protective and sensitizing effects of various compounds and, what is more, molecular radiobiology reveals totally new phenomena which were not exposed in radiochemical experiments but are of crucial importance in the living cell. Thus, molecular radiobiology has discovered specialized enzymatic repair systems which continuously restore spontaneously arising defects in the structure of DNA.

It turns out that the extent of the damage inflicted on DNA macromolecules depends not only on the number of "hits" scored by the ionizing particles but also on the rate of enzymatic repair and the availability of energy for the repair processes. Experimentally revealed factors such as the radiosensitivity of ATP synthesis in the nuclei of some cells and the radiosensitivity of the repair enzymes in others may determine the presence or absence of defects in the DNA macromolecule of the irradiated cell.

The prime objective of molecular radiobiology today is to study in greater detail the repair systems of the cell, the mechanisms involved in the repair of single breaks and various switches of the heterocyclic bases determining the genetic code, and the specific conditions protecting or sensitizing these systems with respect to ionizing radiation. Radiochemical studies of supramolecular native DNA structures, isolated complexes of DNA and hybrid-bound RNA or protein have revealed that these supramolecular structures possess properties of particular interest to radiobiology (Chapters 4 and 5). Molecular radiobiology is also faced with a still more difficult, though essentially solvable problem — that of tracing the radiation-induced changes in the even more intricate complexes composed of chromatin, the molecular structure of which is largely responsible for the manifestation of the information encoded in the DNA molecules.

At the small radiation doses sufficient to kill the organism by causing widespread arrest of cell division or to bring about the death in interphase of the cell population of radiosensitive tissues, the rare penetration of ionizing particles into individual DNA molecules clearly could not be a decisive factor. Radiation-induced blocks in irradiated cell populations arise mainly as a result of disturbances in the regulatory mechanisms responsible for the realization of information, the transcription of DNA, and the repression and derepression of genes (see Chapter 4). Unfortunately, very little is known of the molecular basis of these processes in cells possessing a chromosomal apparatus. Research into the disruptions of these processes will undoubtedly contribute a great deal in this field, just as the discovery of radiation mutagenesis spurred a great advance in general genetics. Studies of the radiosensitivity of gene activation in insect and plant cells have clearly shown that the radiation-induced arrest of gene activation is not associated with the coding system of the macromolecules but rather with the drop in the concentration of the low molecular weight effector as a result of the irradiation. The explanation of this drop remains a fascinating subject for future research.

A closely associated problem concerns the changes in the spectrum of informational or messenger RNA (mRNA) in the irradiated cell. Although research in this field is only beginning, the available experimental procedures (chromatographic fractionation of pulse-labeled mRNA, competitive and saturation hybridization with DNA, etc.) hold great promise.

A general comment on the research objects is worth making here. Phages, viruses and bacteria have become favorite test objects in molecular radiobiology because of their much greater simplicity in comparison with the cells of higher organisms. This approach is quite justified with respect to certain problems. However, the greater complexity of the genetic apparatus of chromosome-bearing cells suggests that these possess a radically different system for controlling the life cycle, quite apart from their dependence on a strictly controlled metabolism; these distinctions may lead to major differences in the course of the radiation-induced processes and consequently in the final effects.

The ultimate purpose of molecular radiobiology is to reveal the molecular mechanisms of radiation-induced damage to higher organisms, including man, with the aim of enabling the control of these processes. This is why research on animal and plant cells both in tissue culture and in vivo should be given particular attention in modern molecular radiobiology despite the greater difficulties involved in such experiments.

The second important field of research in molecular radiobiology concerns the effect of radiation on protein molecules inside the cell and in particular on their synthesis and tertiary and quaternary structure (see Chapter 3). Not surprisingly, attention has been largely focused on a particular category of proteins — the enzymes, the biological activity of which can easily be determined. Judging from radiochemical experiments with pure enzymes exposed to a sufficiently large radiation dose, direct inactivation of enzymes does not appear to be a major factor in the fatal events taking place in the irradiated cell. Moreover, various experiments have shown that the synthesis of polypeptide chains on the ribosomes is quite resistant to radiation in the presence of undamaged mRNA. Yet irradiated cells show drastic changes in the condition of the enzyme protein system.

These changes can be attributed to three major factors. The first relates to the condition of the genes controlling the synthesis of the enzymes examined, whether they are repressed or derepressed. The second factor is the high radiosensitivity of the allosteric functions of enzymes, demonstrated both in our laboratory and by Norwegian workers. We were able to show in the case of aspartokinase and phosphofructokinase that doses having no appreciable effect on the activity of the catalytic center of the enzyme destroy its capacity to react with the effectors which normally reversibly inhibit its function. Consequently, a sharp increase in the activity of such enzymes should occur in the cell, where the enzyme is always to some extent reversibly inhibited by its respective metabolite. This was indeed observed in the case of phosphofructokinase in experiments involving the irradiation of thymocytes.

We believe that the observed high radiosensitivity of the allosteric functions of enzymes is a more widespread phenomenon.

Indeed, one of our current working hypotheses regards the gene repressor as a protein, the properties of which are altered by a low molecular weight, hormone-like effector. More attention is now being devoted to changes in the conformation of proteins associated with the intracellular membranes, changes in the membrane permeability, the appearance of membrane potentials, and other properties. The high radiosensitivity of the allosteric functions of enzymes can be attributed to the effect of radiation on the weak cooperative bonds determining the conformation of the enzyme molecule. Thus, molecular radiobiology is faced with a more universal problem — the effect of radiation on the conformation of protein molecules, where even a minor change may have far-reaching consequences for the actively metabolizing cell. The importance of this problem is also evident from the observed changes in the spectrum of enzymes in the irradiated organism. These observations have revealed radiation-induced damage to the assembly of quaternary protein structures in the irradiated cell.

One of the most urgent problems in modern molecular radiobiology is the radiosensitivity of the processes involved in the assembly of supramolecular structures in the developing cell. Here we arrive at a third field of research — the effect of ionizing radiations on the composition, structure and function of cell membranes (see Chapter 6). The available data on the changes in the permeability of the membranes of lysosomes, mitochondria and nuclei in irradiated cells indicate the importance and promise of future research at the molecular level along these lines.

Speaking of membranes, one should draw attention to the scarcity of knowledge regarding the effect of radiation on biopolymers such as lipoproteins, lipopolysaccharides, and lipopolysaccharide-protein complexes, which are important constituents of the cytoplasmic membranes, plasmalemmae and interstitial substance in the tissues of Metazoa. These substances have attracted attention in connection with recent findings on the regulatory function of the cell membranes and on the changes in these functions as a result of radiation-induced malignant regeneration of the cell. Regrettably, the gaps in our knowledge in this field outnumber the established facts.

This is just a brief outline of the current problems in molecular radiobiology associated with the effect of ionizing radiation on the macromolecules and supramolecular structures constituting the basis of life. Today it is well known that all these macromolecules and structures are "animated" by interactions with low molecular weight cell metabolites combining with these to form a single entity. The remarkable properties inherent in the structure of these macromolecules can only be manifested in the presence of energy-rich molecules, low molecular weight effectors and a large number of metabolites. Molecular radiobiology cannot ignore the rapid reactions taking place in this complex system during irradiation, such as the formation of short-lived free radicals, the outburst and extinction of chain reactions, and the appearance of highly reactive compounds interfering with the metabolism of the cell.

Molecular radiobiology must devote attention to the initial processes taking place in the cell immediately after irradiation processes that play an important, if not decisive role in the manifestation of the radiation effects. Here, molecular radiobiology joins forces with radiation biochemistry and biophysics. This cooperation stems from the very nature of the phenomena under scrutiny. It represents an example of the dialectic logic of the evolution of modern natural sciences.

Only by cooperating with related disciplines will molecular radiobiology be able to make a major contribution toward an understanding of the complex processes taking place in the irradiated cell.

Chapter 1

THE ROLE OF THE PRODUCTS OF WATER RADIOLYSIS IN THE TRANSFORMATION OF BIOPOLYMERS IN AQUEOUS SOLUTIONS

INTRODUCTION

The search for effective means of protecting organisms and cells from radiation necessarily involves the study of the mechanism underlying the first stages of radiation-induced injury of biological objects at the molecular level.

The cell is a complex heterogeneous system with a multitude of components forming solid and liquid phases separated by membranes. Accordingly, the study of radiolytic events in these phases can be focused on the consequences of the direct and indirect effects of radiation in this system.

Here we shall not attempt to give an exhaustive review of the available data in this field of radiobiology, which may be termed radiation biochemistry. A detailed discussion of these topics can be found in several monographs [1—7]. We shall rather focus our attention on one particular problem of radiation biochemistry, namely, the manifestation of the indirect effect of radiation on biopolymers during the early stages of radiation-induced injury, with particular emphasis on research carried out during the last few years.

The radiolysis of aqueous solutions of biopolymers provides a convenient model for the study of the primary stages of radiation-induced injury. There is no doubt that the radiolytic transformations observed in dilute aqueous solutions of various compounds result from their reaction with ionized molecules of water and its dissociation products, namely the free radicals OH and H.

Just as with simple inorganic systems, however, it is more difficult to explain the conversions of solutes induced by the products of the radiolytic breakdown of water in the case of concentrated solutions. Here the possible participation of excited water molecules in the reactions must be taken into account. Such molecules can, for example, transmit their excitation to the solute, react with the latter, or break down into H and OH radicals which in turn may react with the solute [8].

Hence the difficulty of elucidating the role of the radiolytic products of water in the breakdown of biopolymers in solution, i. e., high molecular weight compounds possessing a developed conjugation system, many functional groups and a structure fitting into that of water without excessively distorting it.

In our view, to determine the role of the radiolytic products of water in the breakdown of biopolymers, the major components of the cell, we must first discover the principles of the radiochemistry of aqueous solutions in these biological systems, and secondly trace the fate of the primary products of water radiolysis in these systems, using modern procedures of physicochemical analysis.

THE PRINCIPLES OF THE RADIOCHEMISTRY OF AQUEOUS SOLUTIONS AS APPLIED TO BIOLOGICAL SYSTEMS

The chemical components of the cell as acceptors of the products of water radiolyses. Sites of attack. Reactivity. Protection and sensitization effects

Analysis of the available data on the radiolysis of aqueous solutions of inorganic and organic compounds [8, 9] indicates two general principles.

1. Compounds having different structures and properties behave uniformly with respect to the products of water radiolysis, namely they serve as acceptors of these products during the radiolysis of aqueous solutions.

2. As the solute concentration increases, the reaction gradually involves all the products formed in the radiolysis of water. The relationship between the concentration of the solute and the yield of solute conversion products is expressed by a stepped curve. In physical terms the existence of these steps can be attributed to the fact that totally different types of radiolyzed water molecules may participate in the reaction according to the particular conditions. Different conditions cause the production of ionized and excited water molecules (or their conversion products), differing from one another chiefly in the geometric pattern of their distribution in the irradiated system.

These two general principles provide the basis for the existing methodological approaches to the determination of the function performed by the products of water radiolysis in the conversion of the solute and the mechanism of these conversions.

Information concerning the role of the products of water radiolysis in the transformation of biopolymers constituting chemical components of the cell and their low molecular weight analogs or precursors in the liquid phase, has been obtained by the methods used in the radiochemistry of aqueous solutions. These include indirect procedures, such as measuring the competition between radical acceptors, recording the molecular products of radiolysis — H_2 and H_2O_2 — under different conditions of radiolysis and determination of the conversion products, as well as direct methods, such as kinetic spectrophotometry in conjunction with the pulse method of irradiation, which enables detection of the entry of the primary products of water radiolysis, mainly e_{aq} and OH radicals, into the reaction either by their reaction products or the intermediary products formed.*

---

* This classification of the procedures for recording radiochemical events is, of course, arbitrary.

The method based on competition between radical acceptors as applied to biological objects consists in studying the radiochemical kinetics of the conversion of an indicator solute or its radiolysis product in the presence of different concentrations of the material examined. The indicators used are compounds with a sufficiently well known mechanism of radiochemical conversions; generally the conversion of the indicator is caused by one of the radiolytic products of water. If the tested compound competes with the indicator for the same component of the radiolysis of water, the indicator reaction will be slowed down at certain concentrations of the solute. Apart from showing the particular component of water radiolysis reacting with the tested substance, this procedure enables the determination of the reactivity of the biological material toward the products of water radiolysis in comparison with the reactivity of the indicator using the steady state method.

Knowing the rate constant of the reaction between the indicator and a certain component of water radiolysis, the absolute rate constant of the reaction can be calculated. Naturally, the accuracy of the constants obtained depends on how well the mechanism of radiolysis in the indicator system has been studied.

Such procedures have been used for determining the reactivity of numerous compounds of biological origin and the rate constants of the reactions with the oxidative component of water radiolysis [10−28]. Table 1.1 shows the values of constants taken from the literature [21, 28] and corrected in the light of more accurate determinations of the rate constants of indicator reactions for solutions of different pH and at different concentrations of the indicator compound. The indicator reactions used [20−23, 28] include the oxidation of p-nitrosodimethylaniline (RNO) [10], for which the oxidation rate constant equals $1.25 \cdot 10^{10} \, M^{-1} \cdot sec^{-1}$ [29], and the oxidation of the hydroxylamine disulfonate ion (HADS) to the peroxylamine disulfonate ion (PADS), which is stable in water [30−34]. The absolute rate constant for the reaction of HADS with OH (for solutions of pH 8.8−11) has been calculated from the slope of the linear anamorphosis curve showing the yield of the conversion of HADS (or more precisely, the accumulation of PADS) as a function of the relative concentration of benzene introduced into the solution; $k_{HADS+OH} = (4.9 \pm 1.0) \cdot 10^7 \, M^{-1} \cdot sec^{-1}$. The value $k_{benzine+OH} = (7.8 \pm 1.1) \cdot 10^9 \, M^{-1} \cdot sec^{-1}$, used in this calculation, was determined by the method of pulse radiolysis using solutions at pH = 7 [35].

The rate constant for the oxidation of HADS in solutions of pH ≥ 12 is $k_{HADS+O^-} = 1.43 \, k_{HADS+OH^-}$, since it was found that in solutions with a pH higher than 8.8 the radiochemical yield of the stable radical PADS is 1.43 times greater [34], whereas $pK_{OH} = 11.85$ [36]. The effects obtained with these indicator systems were recorded respectively by means of spectrophotometry and electron paramagnetic resonance (EPR), respectively.

The values for the rate constant shown in Table 1.1 and those published in the literature [10−28] deserve attention for a number of reasons. Firstly, they enable quantitative determination of the role of the oxidative component of water radiolysis, i.e., of its total utilization through reactions with the solute. This is important because in the direct procedure for recording the products (pulse radiolysis combined with spectrophotometry) data on

TABLE 1.1. Rate constants of the reaction between various compounds and OH, determined by means of the HADS + OH (pH = 12) and RNO + OH (pH = 6.5) indicator systems

Compound	$K_{OH+compound}$, $M^{-1} \cdot sec^{-1}$	Compound	$K_{OH+compound}$, $M^{-1} \cdot sec^{-1}$
DNA	$(7\pm1.7) \cdot 10^{10}$	Glucose-1-phosphate-	$(4.2\pm0.7) \cdot 10^{7}$
	$(5.6\pm1.1) \cdot 10^{9}*$	dipotassium salt	$(1.6\pm0.4) \cdot 10^{8}**$
	$(2.2\pm0.4) \cdot 10^{10}**$	Na_2HPO_4	$(1.6\pm0.3) \cdot 10^{8}$
RNA	$(1.9\pm0.3) \cdot 10^{9}**$	NaH_2PO_4	$(8.2\pm0.2) \cdot 10^{7}$
Thymine	$(7.6\pm1.5) \cdot 10^{9}$	Human serum albumin	$(4.0\pm0.8) \cdot 10^{8}*$
	$(4.4\pm0.9) \cdot 10^{9}*$		$(9.4\pm2.2) \cdot 10^{7}**$
	$(1.4\pm0.1) \cdot 10^{9}**$	β-phenylalanine	$(3.7\pm0.7) \cdot 10^{9}$
Thymidine	$(2.7\pm0.5) \cdot 10^{9}$		$(2.8\pm0.6) \cdot 10^{9}**$
Thymidylic acid	$(6.9\pm0.2) \cdot 10^{8}$	Tyrosine	$(1.2\pm0.35) \cdot 10^{10}$
Uracil	$(4.2\pm0.8) \cdot 10^{9}$		$(5.6\pm2.6) \cdot 10^{9}**$
	$(6.9\pm1.3) \cdot 10^{9}**$	Tryptophan	$(4.3\pm0.9) \cdot 10^{10}$
Uridine	$(2.4\pm0.5) \cdot 10^{9}$	Cysteine	$(5.6\pm1.9) \cdot 10^{9}$
	$(2.4\pm0.1) \cdot 10^{9}**$	Cystine	$(5.7\pm0.9) \cdot 10^{10}$
2',3'-Uridylic acid	$(1.9\pm0.5) \cdot 10^{9}$		$(3.1\pm0.3) \cdot 10^{8}$
	$(2.5\pm0.3) \cdot 10^{9}**$		$(4.4\pm0.8) \cdot 10^{9}$
5-Aminouracil	$(8.8\pm1.7) \cdot 10^{9}$		(pH = 11)
Cytosine	$(2.9\pm0.6) \cdot 10^{9}$	Alanine	$(9.8\pm1.9) \cdot 10^{7}$
Cytidine	$(2.5\pm0.4) \cdot 10^{9}$	Glycine	$(5.5\pm1.7) \cdot 10^{8}*$
Deoxycytidine ·HCl	$(6.4\pm2.7) \cdot 10^{9}$		$(3.7\pm1.9) \cdot 10^{8}$
Deoxycytidylic acid	$(8.5\pm2.3) \cdot 10^{9}$	Deoxyphenylalanine	$(3.9\pm0.8) \cdot 10^{9}**$
Adenine	$(4.9\pm1.0) \cdot 10^{9}$	DNP	$(1.36\pm0.3) \cdot 10^{10}$
Hypoxanthine	$(4.2\pm0.8) \cdot 10^{9}$	Benzene	$(2.3\pm0.7) \cdot 10^{9}$
Adenosine	$(2.8\pm0.3) \cdot 10^{9}$		(pH = 11)
Adenylic acid	$(3.6\pm1.4) \cdot 10^{9}$	Phenol	$(2.4\pm0.5) \cdot 10^{10}*$
Deoxyadenosine	$(2.0\pm0.4) \cdot 10^{9}$		$(7.7\pm1.5) \cdot 10^{10}$
Deoxyadenylic acid · NH_4	$(2.0\pm0.7) \cdot 10^{9}$	Indole	$(4.7\pm0.8) \cdot 10^{10}$
Guanine	$(4.1\pm0.3) \cdot 10^{9}$	Maize starch	$(2.9\pm0.5) \cdot 10^{8}**$
Deoxyguanosine	$(1.25\pm0.37) \cdot 10^{9}$	Waxy starch	$(2.5\pm0.4) \cdot 10^{8}**$
	(pH = 11)	Glucose	$(9.5\pm1.9) \cdot 10^{7}$
Deoxyguanylic acid · NH_4	$(2.3\pm0.6) \cdot 10^{9}$		$(3.8\pm1.1) \cdot 10^{8}**$
	$(9.6\pm2.9) \cdot 10^{8}$	Cellobiose	$(4.9\pm1.0) \cdot 10^{8}$
	(pH = 11)		$(3.7\pm0.1) \cdot 10^{9}**$
2-Deoxyribose	$(8.8\pm1.9) \cdot 10^{8}$	Melibiose	$(3.8\pm0.2) \cdot 10^{9}**$
Ribose	$(1.36\pm0.45) \cdot 10^{9}$	Lactose	$(2.4\pm0.2) \cdot 10^{9}**$
	$(4.4\pm0.4) \cdot 10^{8}**$	Methyl glucoside	$(2.4\pm0.3) \cdot 10^{9}**$
		Methyl galactoside	$(1.6\pm0.1) \cdot 10^{9}**$
		Methyl arabinoside	$(2.4\pm0.5) \cdot 10^{9}**$

Note. In determining the relative rate constants of the reactions (G) from the slope of the linear an-amorphosis curves, it was taken into account that the radiochemical yield of OH radicals (or more precisely, the number of OH radicals involved in the reaction) depends on the concentration of the indicator in the solution /32/ so that: G(OH) = 10±2 at [HADS] = 0.1 M, pH = 12; G(OH) = 7±1.5 at [HADS] = 0.1 M, pH = 8.8; G(OH) = 5±1.5 at [HADS] = 0.3 M, pH = 8.8; G(OH) = 2.6 at [HADS] = $5 \cdot 10^{-3}$ M, pH = 8.8.

* Values of the constant obtained with the HADS+OH indicator system.
** Values of the constant obtained with the RNO+OH indicator system.

the participation of the oxidative components of water radiolysis are derived solely from the accumulation of an intermediary product absorbing light in a certain part of the spectrum (a product formed with the addition of OH radicals, apparently for the most part at double bonds) [37–40]. The loss of OH or O⁻ radicals cannot be recorded directly by spectrophotometry because their absorption bands (220–280 nm) [41] may overlap with those of the initial objects or their radiolysis products.

The radiobiological importance of the oxidative component during the primary stages of radiolysis can hardly be overestimated since the radiolysis of cell components under natural conditions takes place in the presence of a large amount of molecular oxygen (about 10^{-3} M). Such a system does not permit direct reductive reactions either with e_{aq} or with H atoms, since molecular oxygen transforms e and H into O_2^- and HO_2, respectively, i.e., into the oxidative component of radiolysis: $k_{O_2 + e_{aq}} = 1.88 \cdot 10^{10} \, M^{-1} \cdot sec^{-1}$ [42], $k_{O_2 + H} = 2.1 \cdot 10^{10} \, M^{-1} \cdot sec^{-1}$ [43]. Admittedly, in reactions in which O_2 serves as an electron carrier the electron hits the object affected.

Secondly, the rate constants for the reaction obtained with different indicator systems in principle enable a leveling out of the discrepancies existing between different data probably as a result of the specific conditions of radiolysis or the effect of intermediary transformation products of the indicator or test compound in solution.

Thirdly, it is noteworthy that the rate constants for the reaction obtained with comparatively concentrated solutions (i.e., under conditions in which the conversion products of both ionized and excited water molecules participate in the reaction) are practically identical to those obtained with dilute solutions of the same compounds. This means that the same reactive particle — the OH radical — participates in reactions with the solute in concentrated and dilute solutions alike. Two conclusions can be drawn from this fact: 1) excited water molecules do not enter the reaction as such but break down into radicals; 2) the reaction of the OH radicals with the solute is preceded by a faster stage — the transfer of excitation energy to the water molecules included in the hydrate sheaths of the functional groups of the solute.

Comparison of the data obtained using various indicator systems leads to the following conclusions.

1. The biopolymers examined and their precursors behave as typical acceptors of OH radicals.

2. DNA, proteins, DNP and polysaccharides vary in their reactivity with respect to OH radicals.

In evaluating the reactivity of biopolymers, however, one must bear in mind that the rate constants for reactions involving biopolymers, as determined in indicator systems, are higher than those obtained for their low molecular weight analogs. Some workers explain this discrepancy by assuming that the radiolysis of biopolymers creates conditions conducive to a direct attack on particular functional groups of the biopolymer (in other words, that the active product can "see" these functional groups better in the biopolymer [16]). The indicator compound possibly loses some of its motility in the solution of the biopolymer, either because of the structure of the latter or because it binds with certain functional groups of the biopolymer.

Another factor to consider when determining the reactivity of biopolymers and their low molecular weight analogs is that values calculated on the basis of the reactivity of the individual components of the biopolymer (for example, in the case of solutions of human serum albumin [23]) exceed those actually measured experimentally. The probable reason for this is that the OH radicals in solutions of the precursors react mainly with the functional groups of the precursors involved in the synthesis of the biopolymer, or, in other words, the structure of the biopolymer impedes the access of OH to the reacting functional groups.

The data presented in Table 1.1 and other published data indicate the most probable sites of attack by OH radicals in the molecules of biopolymers in protein molecules, for example, the OH radicals attack preferentially the sulfur-containing and aromatic amino acid residues; in the latter, the aromatic rings are most vulnerable [23]. (This applies generally to reactions of H atoms [24].) In the case of nucleic acid molecules, determinations of the reactivity of the different components — bases, 2-deoxyribose and the phosphate ion — show that about 80—90% of the OH radicals attack the bases and only 10—20% the sugar-phosphate fragment.

As a rule, radio-protective compounds are highly reactive toward the oxidative component of the radiolysis of water in experiments involving competition between acceptors (Table 1.2). The protective effect of such compounds in biological systems may be due, at least in part, to their competition with the biopolymer for the oxidative component of water radiolysis.

TABLE 1.2. Rate constants of the reaction between radioprotectors and the oxidative and reductive components of the radiolysis of water

Radioprotector	$k_{compound + OH}$ $10^9 M^{-1} \cdot sec^{-1}$	$k_{compound + e_{aq}}$ $M^{-1} \cdot sec^{-1}$	References
Mesoinositol	1.0±0.12	—	[21, corrected by us]
Propyl gallate	12.0±2	$(5.5±2.2) \cdot 10^5$	[21,28]
3-Oxypyridine	6.7±0.6	—	[21]
2,4,6-Trimethyl-3-oxypyridine	2.5±0.4	$(2.0±0.2) \cdot 10^7$	[21,28]
4-Ethyl-6-methyl-3-oxypyridine	1.4±0.1	—	[21]
dl-Penicillamine	4.7	—	[44]
Penicillamine disulfide	6.5	—	[44]
2-Mercaptoethylamine	14.9	—	[44]
2,2'-Dithio-bis-ethylamine	11.0	—	[44]

A further conclusion which can be drawn from Table 1.1 is that the dissolved low molecular weight precursors of the biopolymers, present in a free state in the cytoplasm, may serve as natural radioprotectors of the biopolymers themselves; this supposition is confirmed by analysis of the final products of the radiolysis of the system DNA—phenylalanine—water in deamination processes [45].

The direct recording of the molecular products of the breakdown of water — H_2 and H_2O_2 — in aqueous solutions of biological objects provides data on the reactions of these products under conditions which exclude the influence of secondary processes and other complications such as reactions with the transformation products of the indicator compounds, characteristic of the method discussed above.

This method is based on the assumption that the components of water radiolysis recombine into H_2 and H_2O_2 if they do not react with the solute. In neutral solutions these products can be formed by the following reactions:

$$H + H = H_2; \tag{1.1}$$

$$e_{aq} + e_{aq} + 2H^+ = H_2 + H_2O; \tag{1.2}$$

$$OH + OH = H_2O_2; \tag{1.3}$$

$$O_2^- + O_2^- + 2H^+ = H_2O_2 + O_2 \tag{1.4}$$

(1.4 in the presence of O_2). It is noteworthy that in this case the effect of concentration on the yield G_{mp} of molecular products of the radiolysis of water can be expressed as $G_{mp} \approx G_{mp}^0 - B[Ac]^{1/3}$ which reflects the competitive processes involved in the recombination of the radicals formed along the track of the ionizing particles [8, 9, 34, 46–53]. If the concentration [Ac] of the acceptor is extrapolated to zero, the intersect of the straight line on the ordinate axis (with the coordinates $G \sim [Ac]^{1/3}$) shows the maximum value of the yield G_{mp}^0 of the molecular products H_2 and H_2O_2 in the ideal case in which all the H and OH radicals formed along the track recombine only with radicals of the same kind. For H_2 this value is 0.45.

The scant available data on the effect of concentration on the yield of radiolytic products suggest that in aqueous solutions of biological objects a linear relationship will be obtained not only for the yield of the molecular products of water radiolysis but also for the breakdown of the initial compounds and the accumulation of their radiolytic products [52, 53]. This is the case when the OH and H radicals react predominantly with double bonds, for example, those contained in the aromatic rings of amino acids. In processes of the type

$$RH + H = R + H_2, \tag{1.5}$$

the effect of concentration should be similar but the maximum yield of H_2 should increase to at least twice the value obtained in the former case. Measurements of the yield of molecular hydrogen in the radiolysis of aqueous solutions of glycine [54] and glucose [19] have shown that such processes can take place in biopolymers containing aliphatic amino acid residues and monosaccharide units. The yield of H_2 in the radiolysis of aqueous solutions of thymine at different concentrations does not exceed 0.46. In a 0.03% solution (by weight) of DNA $G(H_2) = 0.37$ [55]. This means, firstly, that e_{aq} is almost completely taken up by the substrate in solutions of DNA at this concentration, and, secondly, that atomic hydrogen in DNA solutions practically does not participate in the reaction (1.5) but reacts with double bonds.

In studying the effect of concentration on the yield of H_2O_2 in aqueous solutions of biological substances one must bear in mind that the kinetics of the accumulation of H_2O_2 may be complicated by reactions leading to the formation of H_2O_2 as a radiolytic product of the solute, for example, according to the reaction sequence $R \rightarrow RO_2 \rightarrow ROOH$, with the subsequent breakdown of the hydroperoxide compound and the release of hydrogen peroxide. The newly formed H_2O_2 may also react with the products of radiolysis or with the original compounds. All these processes are responsible for the complex kinetics of the formation and consumption of H_2O_2 in the radiolysis of aqueous solutions of carbohydrates.

Thus, the reactions involving the products of water radiolysis (e_{aq}, H, and OH) during the radiolysis of aqueous solutions of biological systems can often be identified by determining radiochemically the effect of concentration on the accumulation of the molecular products of water radiolysis.

Analysis of the products of radiolysis constitutes a direct method for determining the participation of the products of water radiolysis in the reactions. Thus, chromatography reveals the products obtained by the addition of OH at double bonds in nucleotide bases [56–58], phenylalanine (tyrosine), and tyrosine (deoxyphenylalanine). Chemical and physico-chemical analysis of the composition of the radiolytic products of nucleic acids and low molecular weight polysaccharide analogs has led to the conclusion that the OH radicals attack mainly the C'_1, C'_3, and C'_5 atoms of the sugar component of DNA (see below for further details). This method in conjunction with that of recording the primary products of the radiolysis of biological systems may ultimately enable a thorough study of the mechanism of the transformation of cell components under the influence of radiation.

The method of direct recording of the primary products of water radiolysis and tracing their reactions during irradiation has largely been worked out; it applies to reactions involving the hydrated electron e_{aq}, since this particle can be reliably detected by its absorption spectrum in the range of visible light. Numerous studies have been devoted to determining the reactivity of e_{aq} with respect to biological objects on the basis of the reaction rate constants. The largest collection of data, covering about 100 biological compounds, may be found in the monograph by Pikaev [59]. Of the biopolymers examined (proteins, DNA, polysaccharides), the protein molecule shows the highest reaction rate with respect to e_{aq}. The rate constant of the reaction with e_{aq} is much lower in the case of DNA and polysaccharides such as hyaluronic acid [60] and starch [28]. It is noteworthy that of all the amino acids studied, only those containing sulfur (cystine, cysteine) are highly reactive toward e_{aq}; the aromatic amino acids do not differ in this respect from the aliphatic ones [40, 61, 62]. This suggests that e_{aq} attacks the sulfur-containing amino acids in the protein molecule.

The nitrogen bases within nucleic acid molecules similarly show a high reactivity toward e_{aq}. Radioprotective compounds containing sulfhydryl and $-S-S-$ groups are active acceptors of e_{aq} [61–63].

Using procedures such as pulse irradiation and kinetic spectrophotometry, it is possible to study the involvement of e_{aq} and OH in reactions with the substrate and also to trace the transformations of the primary products such as the transfer of electrons to radioprotective or radiosensitizing compounds [64, 65], or the interaction between the radicals formed from the substrate and additives, such as stable radicals, introduced into the solution [66, 67]. The rate constants of these processes may be determined at the same time. It was found [64, 65] that the rate constants for the transfer of electrons from DNA components to certain radioprotectors are close to the diffusion constants. These processes presumably play a major role in native systems.

These findings suggest two possible modes of action of radioprotective and radiosensitizing compounds. One possibility is that they compete with the substrate for the radiolytic products e_{aq} and OH, yielding a little active product in the case of a radioprotector and one or more active products in the case of a radiosensitizer. Alternatively, they may capture the unpaired electron from the initially formed product or react with a radical formed from the substrate.

Effect of concentration on the yield of solute conversion. Sensitization effects

The main parameter of the radiochemical conversion of any solute is the value of the radiochemical yield G. Depending on the conditions of radiolysis, this value ranges from thousandths to 10 or more transformed molecules or ions per 100 eV of absorbed energy for the same compound in aqueous solution [8, 68–72]. The degree of radiolysis of the solute depends, firstly, on the number of radiolyzed water molecules participating in the reaction (the maximum number being 12 [8, 68–72]) and, secondly, on the reactivity of the intermediate radiolytic products of the solute in relation to that of the original compound. Naturally, each of these basic factors affecting the yield G may be masked by side processes and reverse reactions. It may therefore be difficult to explain the high yield of one radiation-induced process and the low yield of another during radiolysis under natural conditions in the cell.

Nevertheless it is essential to know the basic factor determining the breakdown of a particular cell component, as well as the maximum and minimum yields of this breakdown (under conditions of sensitization and protection, respectively) in view of the practical need to regulate the processes involved, either toward maximum breakdown of the component (radiosensitization therapy) or toward maximum suppression of the radiation-induced damage (in the search for the most effective protection). Below we shall try to identify and analyze the causes of the changes in the radiochemical yield of conversions of the main components of the cell. First of all let us consider the manner in which the primary products of water radiolysis enter the reaction. The basic conditions for the involvement of the products of water radiolysis in the reaction have been determined in Proskurnin's laboratory [8].

1. With a rise in the solute concentration there is a gradual increase in the involvement of the active products of water radiolysis – at first ionized products (about 4) or their conversion derivatives, and afterwards excited products or their conversion derivatives – H and OH radicals (likewise about 4) – owing to inhibition of the recombination of radicals in the cell: $H + OH = H_2O$. This is reflected in the stepwise course of the curves showing the effect of concentration on the yield of radiolytic products, corresponding to the total involvement of the different products of water radiolysis in

the reaction with increasing concentration [8, 71–73]. The maximum yield of solute conversion achieved in this manner is 8–9.

The geometric course of the track changes with time owing to the diffusion of active particles, their recombination with each other and their reactions with the solutes. The observed concentration curve of the yield reflects the final state of the reactions occurring along the track, the proportion of the active products of water radiolysis reacting with the solute in comparison with the total quantity of active products participating in reactions. In the series of such curves obtained, that situated farthest to the left corresponds to the most reactive compound [72].

2. Water may be regarded as a relatively highly structured liquid. The energy absorbed in the form of excitation can migrate through the system along channels created by the structure of the liquid; there is a distinct probability that the excitation energy will be concentrated at molecules located at sites where the water structure is distorted, i.e., in the hydrate envelopes of solutes [71]. Factors such as the degree of interaction between the excited water molecules closest to the solute and the solute itself (their bonding), and the reactivity of the solute toward H and OH radicals, ultimately determine whether the excitation energy will be transferred to the solute or whether the water molecule will break down into H and OH radicals and these react with the solute.

3. The breakdown of the water molecule into radicals and the involvement of these in reactions are enhanced by the simultaneous radiolysis of two solutes, each of which predominantly binds a different component of the radiolysis of water, i.e., the two compounds act as conjugated radical acceptors (according to Proskurnin). In the presence of conjugated acceptors conversion reaches a peak at lower concentrations. Conjugated acceptors can operate in comparatively dilute solutions, where the solute molecules are far apart. This long-range effect may be associated with a mechanism involving the migration of the excitation along a chain of water molecules held together by hydrogen bonds and terminating with the conjugated acceptors [9, 69] which thus accept radicals generated from the water by a relay process. Analogous to this is the introduction of excited water molecules into the reaction in the presence of substances possessing a great affinity for protons and consequently altering the chemical nature of the radiolyzed particles:

$$OH + \xrightarrow{+ OH^-} O^- \cdot H_2O$$

$$H \xrightarrow{+ OH^-} e \cdot H_2O, \qquad (1.6)$$

as is the case with strongly alkaline solutions. The maximum yield of solute conversion by these mechanisms is 12.

All these conditions may, in principle, be implemented during the irradiation of areas of the cell constituting a liquid phase, an aqueous solution, or more precisely, several solutions ranging from very dilute to highly concentrated (including gels) and possessing a set of functional groups which can be mutually competitive or act as conjugated acceptors.

On the basis of these considerations we shall try to analyze the available data on the radiolysis of aqueous solutions of the chemical components of the cell. This analysis will deal in particular with effects detected during the primary stages of radiolysis before the onset of subsequent processes.

The most complete account of the effect of concentration on the conversion yield of dissolved cell components in aqueous solutions can be found in the monograph by Amiragova et al. [4]. According to this, in the radiolysis of low molecular weight precursors of nucleic acids (nucleotides, nucleosides, nitrogen bases), the curves showing the effect of concentration on the yield of conversion products from the breakdown form a distinct plateau in the concentration range reaching 10^{-2} M. The maximum yield of the radiolytic breakdown of nucleic acid components (as determined by the rupture of the double bonds in the chromophore groups) in the presence of oxygen is about 5 per 100 eV of absorbed energy.

In the radiolysis of protein precursors such as aromatic (4, 51—53, 75] and sulfur-containing [76—81] amino acids (such data are of particular interest since amino acid residues are the main target of attack by OH and e_{aq} radicals) and reduced glutathione the breakdown yield increases with the concentration and the curves similarly tend to approach saturation or reach a plateau at concentrations ranging from 10^{-4} to 10^{-1} M. In the case of radiolysis of aqueous solutions of tryptophan, for example, at concentrations ranging from $2.5 \cdot 10^{-5}$ to 10^{-3} M, the effect of concentration on the tryptophan breakdown and the accumulation of radiolytic products is characterized by typical saturation curves [4, p. 252]. The maximum yield of tryptophan breakdown in a $2.5 \cdot 10^{-4}$ M solution is 3.3 [4, pp. 75, 254]. The maximum yield of the breakdown of phenylalanine in solutions (in the absence of O_2) is 10 (in a 0.1 M neutral solution [53]); of tyrosine — 5.7 ($1.5 \cdot 10^{-1}$ M in a sulfuric acid solution [80]); of cystine — 3.5 ($5 \cdot 10^{-4}$ M [77]); the radiochemical oxidation yield of cysteine hydrochloride is 9.24 ($5.1 \cdot 10^{-2}$ M [78, 81]). The breakdown yields of the initial compounds in the radiolysis of methionine solutions and in the reduction of glutathione in the absence of oxygen are respectively 4.7 ($5 \cdot 10^{-2}$ M [79]) and 8.1 (10^{-3} M, pH = 5.3 [82]). Of particular interest is the yield of the breakdown of histidine [4, p. 255] and the accumulation of ammonia in serine and glycine solutions [4, p. 221] at the highest concentrations used (on the plateau of the concentration curves) — more than 10—12 molecules per 100 eV of absorbed energy.

In the radiolysis of low molecular weight analogs of polysaccharides — monosaccharides and disaccharides of varying structure — the effect of concentration on the radiolytic breakdown in aqueous solutions gives typical saturation curves at concentrations by saturation ranging from $5 \cdot 10^{-5}$ to 1 M. The maximum yield of the breakdown of carbohydrates in oxygen-free solutions varies within the limits of 3 to 4 molecules per 100 eV of absorbed energy [83—85].

The radiolysis of aqueous solutions of biopolymers has not been studied in sufficient detail over a wide range of concentrations. The most complete accounts of the radiolysis of biopolymers in solutions at different concentrations are given in the publications by Amiragova et al. and Krushinskaya [4, 74]. According to these studies the breakdown of DNA (as determined by the destruction of the chromophore groups), as a function of its concentration, for example, shows a logarithmic relationship with a peak of DNA breakdown corresponding to 2—3.

Numerous publications deal with the effect of concentration on the radiolysis of aqueous solutions of proteins, notably enzymes. The main purpose of these studies was generally to discover whether the fundamental property (i. e., the biological activity) of the biopolymers in aqueous solution changes as a result of the action of the radiolytic products of water or of the direct action of radiation on the substrate. These studies [1, 2, 86—103] lead to the following conclusions.

1. The property of the macromolecules recorded — loss of biological activity (i. e., the overall effect of all the changes in the macromolecule induced by radiation impairing its basic biological function) — is subject to one of the basic rules of the radiochemistry of aqueous solutions [97], i. e., that the concentration curve of G (inactivation) attains a plateau at the highest concentrations of the substrate. This proves that products of the radiolysis of water participate in the reaction.

2. The radiochemical yield of enzyme inactivation depends largely on the nature of the enzyme and ranges from 0.009 to 0.55 [1, 2, 86—92, 97].

3. Enzyme inactivation is caused mainly by OH and H radicals [1, 90, 95, 98], which attack tryptophan, tyrosine, and sulfur-containing amino acid residues [93—96].

4. Studies of the effect of the radiation dose on the loss of enzyme activity [92] show that the observed deviation from the exponential relationship results from two factors. First of all, the amino acid residues of the enzyme (the work was done with α-amylase) differ in their reactivity toward the products of water radiolysis and the most reactive residues are preferentially "knocked out." (These findings agree with results obtained by the pulse radiolysis method with lysozyme solutions [94].) Secondly, the conformation of the enzyme changes during irradiation (e. g., it becomes denatured).

In many experiments to determine the effect of concentration on the physicochemical properties of aqueous solutions of proteins and enzymes, reflecting the primary stages of radiolysis or stages close to these, it was found that the curves of G as a function of concentration were typical saturation curves [98, 100]. This was the case in processes such as the oxidation of oxyhemoglobin to methemoglobin ($Fe^{2+} \rightarrow Fe^{3+}$) in aerated solutions [98]; the reduction of ferrocytochrome C in deaerated solutions [100]; the reduction of ferricyanide ions, in 20% solutions by weight in gelatin gels [101], in the presence of various gases (N_2, O_2, N_2O, air); the aggregation (cross linking) of macromolecules in collagen solutions at concentrations of 0.055—0.5% by weight [102]. At the highest protein concentrations examined (or on the plateau) the values of G corresponded to a total involvement in the reaction of 3—4 pairs of radicals originating from the water.

Studies of the radiolytic behavior of other hemoproteins in aqueous solution (myoglobin, peroxidase, catalase) reveal that the observed effects in the coordinates of G (for the hemoprotein) are subject to different laws. In aqueous solutions of catalase, for example, the inactivation yield at concentrations ranging from 10^{-9} to $5 \cdot 10^{-7}$ M changes almost exponentially (the process being dependent on the dose of radiation [103]). However, these concentration effects, gauged from the observed properties of the products, may be considerably complicated by the occurrence of multi-stage processes (chain reactions in the case of catalase), even though the primary, triggering processes, resulting from the participation of the radiolytic products of water, are essentially the same as in the former cases. Thorough analysis of the mechanism of radiolysis of these systems enables this assumption to be verified.

It is evident from these data on the effect of concentration on the breakdown yield that low molecular weight precursors or biopolymers with a solubility not higher than 10^{-2} M (calculated for the monomer unit of the polymer) have a maximum conversion yield of 4.5–5 on the plateau of the curve. It is worth noting that this value is obtained precisely when the destruction of one molecule of the original substance requires one OH radical. The value 4.5 corresponds exactly (within 10% limits) to the number of radical pairs formed from the ionized water molecules and from a proportion of the excited molecules (those overcoming the cage effect), that usually react in comparatively dilute solutions where reverse processes are inhibited [73]. Data obtained using the method of electron paramagnetic resonance in frozen, glassy, aqueous solutions of alkalies show that the total yield of the oxidative component of water radiolysis, i.e., the O⁻ radicals (which participate instead of OH in reactions under alkaline conditions), is 4.2 ± 0.2: $G(e_{st}) = G(O^-) = 3.4 \pm 0.2$ from ionized molecules and 0.8 from radiolyzed molecules, which apparently break down into H and OH radicals [104].

If G values smaller than 4.5 are found on the plateau of the curve showing the effect of concentration on the transformation of the chemical components of the cell, it is evident that either processes consuming the products of water radiolysis are taking place in the system, or the breakdown of one molecule of the original compound requires not one but two or more radicals from the water; in this case the value of G is underestimated." An "overestimation" of G, with values greater than 4.5, can result either from the involvement of an additional number of radiolyzed water molecules in the reaction, or from the participation of intermediate radiolytic products in the breakdown of the compound. Thus, in the radiolysis of mono- and di-saccharides, which involves mainly OH radicals,

$$RH + OH = R + H_2O, \tag{1.7}$$

the maximum radiochemical breakdown yield in 0.1–1 M solutions does not exceed 3–4 molecules [83–85]. Since the primary radicals formed from the monosaccharide then disproportionate into the initial compound and another product, at a concentration of about 0.01 M at least all the transformation products of the ionized water molecules must enter into reaction with the sugar [83–85]. In the radiolysis of lactose solutions at a wide range of concentrations ($1 \cdot 10^{-4}$ to 0.67 M), for example, it was found that the value of G (for the lactose) gradually increases with the concentration of lactose in the solution and reaches a peak of 4.27 ± 0.44 [85]. It is noteworthy that the concentration curve of the accumulated yield of one of the more thoroughly studied categories of radiolytic products of lactose — deoxysugars — has a typical stepped course with a plateau at 10^{-3}– $8 \cdot 10^{-2}$ M and at concentrations greater than 10^{-1} M. In our view, this is an indication of the entry of different radiolytic products of water, ionized and excited, into the reaction.

Another interesting fact is that the G values on the "steps" differ from one another by a factor of approximately 2, as is the case with the radiolysis of inorganic compounds over a wide range of concentrations [73]. Since $G(H_2O_{ionized}) \approx G(H_2O_{excited}) = 4$, and the deposition of OH radicals in the formation of deoxysaccharides in centimolar solutions (first step) of lactose saturated with an inert gas is expressed by a G value equal to 0.14, it is evident that at higher lactose concentrations where the reaction involves not e_{aq} ions and OH radicals but only H and OH radicals (from excited water molecules), each of which reacts with the disaccharide (reactions (1.5) and (1.7)) to form the same product, the value of G (deoxysaccharide) must increase to 0.53.

$$G(\text{deoxysaccharide}) \approx G(\text{deoxysaccharide})_{e_{aq}} + G(\text{deoxysaccharide})_{OH_{ionized}} +$$
$$+ G(\text{deoxysaccharide})_{H_{excited}} + G(\text{deoxysaccharide})_{OH_{excited}} = 0.11 + 0.14 +$$
$$+ 0.14 + 0.14 = 0.53.$$

Here the subscript indicates which particles from the radiolyzed water molecules participate in the formation of the deoxysaccharides.

The yield of lactose breakdown observed in concentrated solutions is likewise twice the G (lactose) value on the first step. Comparison of the values of G (carbohydrate) for various kinds of carbohydrates at concentrations corresponding to the first step shows differences from one sugar to another. Thus, the yield depends on the conditions of the radiolysis, i.e., on the involvement of radiolyzed water molecules in the reaction, which in turn is determined by the structure of the radiolyzed solution and the reactivity of the solute. The maximum yield of sugar breakdown observed when a number of compounds were tested in this concentration range was 3.3, obtained with centimolar solutions of cellobiose [85]. Since the participation of e_{aq} in these systems is negligible and G (cellobiose) = 6.5 in centimolar cellobiose solutions saturated with nitrous oxide, we arrive at the following conclusions. Firstly, the value of G (cellobiose) obtained at the first step corresponds exactly to the total participation of OH radicals originating from ionized water molecules, i.e., 3.4 per 100 eV of energy absorbed by the solution [104]; secondly, the radiolysis of sugars (in particular di- and monosaccharides) consists essentially in the removal of an H atom from the molecule and the subsequent transformation of the radical formed from the sugar.

Further examples of the participation of the transformation products of ionized and excited water molecules in reactions with organic solutes can be found in the radiolysis of glycyltryptophan and glucuronic acid and the transformation of acrylamide in aqueous solutions; the mechanism of all these processes has been studied in some detail [105–108]. The radiolysis of glycyltryptophan was studied by the chemiluminescence method. It was found that the luminescence of the irradiated solutions (in the presence of O_2) results from the transformation of the aromatic residue of tryptophan, by the action of OH radicals. The intensity of luminescence of irradiated glycyltryptophan solutions increases with the concentration, at first gradually from 10^{-7} to $3 \cdot 10^{-3}$ M, then rapidly, from a concentration of about $3 \cdot 10^{-4}$ M onward. The curve showing the effect of concentration plotted on semilogarithmic coordinates, has a characteristic break in this area. The curve of luminescence as a function of concentration does not differ essentially from that showing the effect of concentration on the yield of transformation products as determined in the radiolysis of aqueous solutions of inorganic compounds [73]. At higher concentrations the maximum intensity of luminescence is about 3 times greater than that obtained in dilute solutions of glycyltryptophan. This effect may be explained by assuming that the breakdown of the tryptophan ring in more concentrated solutions involves OH radicals not only from ionized water molecules but also from excited water molecules. Since the yield of ionized water molecules is about half the yield of excited water molecules (of both kinds [72]), it follows that the radiochemical effect must be tripled in concentrated solutions, where the OH and H radicals from these water molecules participate in reactions with the solute (OH radicals with glycyltryptophan and H atoms with O_2).

These examples suggest that aqueous solutions of biological and inorganic compounds alike conform to the principles applying to the radiolysis of aqueous solutions generally. In other words, the products of water radiolysis play a dominant role in these systems.

Sensitization of the radiolysis of solutes at different concentrations can be achieved by the introduction of additives into the solution. In this case there are two plausible mechanisms for the action of the additive. Firstly, it may transform a radiolytic component of water essentially inactive toward the solute into one which is active and functions in the same way as a nontransformed active component of water radiolysis. Under such circumstances the breakdown yield of the tested solute at the same concentration but in the presence of the additive, must be doubled and reach 8.4—9 molecules per 100 eV of absorbed energy $[(4.2 + 4.5) \cdot 2]$ at the first step and about 18 and 24 at the second and third steps, respectively. The second possible mechanism of sensitization by additives involves the conjugated action of radical acceptors: the additive causes the breakdown of excited water molecules into radicals and the participation of these in the reaction. As in the former case, the breakdown yield under such circumstances will be more than 8—9 at the first step, more than 18 at the second step and, naturally, 24 in the extreme case.

The first mechanism operates, for example, in the radiolysis of 0.01 M solutions of DNA, lactose and other carbohydrates saturated with nitrous oxide. Studies of the inactivation of DNA in aqueous solutions over a wide range of concentrations (10^{-4}—0.1% by weight) have shown that the process is caused by OH radicals. The saturation of DNA solutions (about 0.1% by weight) with nitrous oxide — an effective transformer of the reductive component into an oxidative one according to the reaction $N_2O + e_{aq} + H^+ = N_2 + OH$, $k_{N_2O + e_{aq}} = 8.67 \cdot 10^9 M^{-1} \cdot sec^{-1}$ [42] — doubles the yield of DNA inactivation in comparison with nitrogen-saturated solutions ($G_{inact} = 1.9 \cdot 2 = 3.8$) [109]. In the case of 0.01 M solutions of low molecular weight polysaccharide analogs — mono- and disaccharides — which are likewise radiolyzed mainly by OH radicals, saturation of the solution with nitrous oxide increases the breakdown yield to approximately twice that obtained from nitrogen-saturated solutions, i. e., to about 4.3 [84, 85].

Some data have been published on the sensitizing effect of molecular oxygen, but unfortunately only a few studies deal with the effect of oxygen on the radiochemical breakdown of biopolymers and their precursors over a wide range of concentrations. Molecular oxygen is an efficient acceptor [42] and can act as a N_2O-type sensitizer if the O_2^- and HO_2 radicals formed in the process, as well as the OH radicals, are aggressive with respect to the material examined. This may be the case in the radiolysis of comparatively dilute solutions of nucleotides involving the rupture of sugar-phosphate bonds (oxygen effect ~ 2 [101]); in the breakdown of phenylalanine, G (phenylalanine) in the presence of O_2 in centimolar neutral solutions is 6.6 ± 0.5 [51, 52].

The oxygen effect is also evident in the radiolysis of sulfur-containing compounds (amino acids, peptides, and some radioprotectors). In highly concentrated solutions of cysteine $(5.1 \cdot 10^{-2} M$ [78, 81]), for example, the breakdown yield changes from 9.26 to 73.92 with the transition to solutions saturated with air; the corresponding increases for methionine solutions [79] at a concentration of $10^{-3} - 5 \cdot 10^{-2} M$ and for glutathione solutions at a concentration of $4 \cdot 10^{-5} M$ [111] are from 4.7 to 9.7 and from 2 to 6.65, respectively. Studies of the radiolysis of compounds containing an $-S-S-$ group have shown that the presence of oxygen does not affect their breakdown yield [78, 81]. The greater breakdown yield of cysteine in highly concentrated solutions in the presence of O_2 may be attributed to chain reactions. In the radiolysis of dilute solutions of cysteine, $10^{-4} M$, where the breakdown of the amino acid by chain reaction is greatly hindered (in comparison with $5.1 \cdot 10^{-2} M$ solutions, for example), the oxygen effect at pH 8 is less pronounced and the maximum transformation yield attains a value approximately equal to about twice the number of radical pairs, or the number of water molecules fully involved in the reaction, i. e., 12.

It is noteworthy that when the sodium salt of S-(2-aminoethyl)-isothiuronium, a radioprotector [112] examined over a comparatively wide range of concentrations $(10^{-5} - 10^{-2} M)$ [113], is radiolyzed in the presence of oxygen, two steps appear in the concentration/yield curve, at 10^{-4} and $10^{-2} M$. The breakdown yield at these concentrations is respectively 9 and 17.1, which, generally speaking, corresponds to the involvement of the transformation products of the ionized water molecules and part (0.5) of the excited ones, assuming that HO_2 or O_2 has one oxidative equivalent in the reaction with this compound, as is the case with OH.

If this conclusion is correct it is possible, as in [9], to determine the average maximum range of action of the conjugated acceptors. The magnitude of the long-range effect depends on the reactivity of the radical acceptors. In solutions of cysteine and oxygen, for example, the distance between the conjugated acceptors (at O_2 concentrations of about $10^{-3} M$) must not exceed 160 Å between cysteine molecules and approximately 120 Å between O_2 molecules. In the radiolysis of S-(2-aminoethyl)-isothiuronium the distance between the conjugated radical acceptors is about 120 Å. Judging from the effect of concentration on the radiolysis of some compounds, the greater radiolytic yield obtained in solutions of low molecular weight compounds or biopolymers at concentrations greater than 10^{-3} and 10^{-2} M (on the basis of the monomer units) may be due not only to the involvement of excited water molecules in the process but also to reactions between the active intermediate transformation products of one solute and the other initial compounds. Such conditions arise during the simultaneous radiolysis of phenylalanine and glycine in aqueous solutions, where radicals from glycine participate in the breakdown of phenylalanine [52]. The yield of the breakdown of cytosine with the formation of ammonia is doubled if formic acid or ethanol is introduced into the solution and remains constant (at 2.9 and 2.3) in the 0.5−0.65 M concentration range, since the radicals originating from the additives react with the cytosine in the same way as the OH radicals [114].

Processes involving the formation of hydroperoxide radicals and hydroperoxides and the breakdown of these may occur in the presence of O_2 in comparatively concentrated solutions of low molecular weight compounds as well as in biopolymer solutions. These processes were detected by the chemiluminescence method in the radiolysis of protein [105] and nucleic acid [115] solutions. The high yield of the breakdown of biopolymers, accompanied by DNA deamination, is not determined by the radiolytic stages but rather by the instability of the radiolytic products − the hydroperoxides [116, 117].

Thus, if the processes obviously do not involve chain reactions and occur in the presence of molecular oxygen, the maximum value of the breakdown yield G does not exceed 8−9 in solutions of sulfur-containing amino acids and peptides [82], or 18 in solutions of radioprotectors [113]. This means that the radiolytic processes can be adequately explained on the basis of the complete utilization of 4.5 and 9 radiolyzed water molecules (the doubled values on the steps of the concentration curve), while the molecular oxygen acts in conjugation with OH, both serving as oxidative agents.

Apart from the sensitizing effect of molecular oxygen in solutions subjected to radiolysis, the substrate may be oxidized by the oxygen generated within the system in a singlet excited state [118]. This aspect, however, is still poorly understood; its clarification requires much more experimental evidence.

The procedures outlined for the analysis of the events occurring during the radiolytic breakdown of the chemical components of the cell can also be applied in the search for effective means of protecting the cell against ionizing radiations. Thus, radioprotectors must be effective receptors of OH radicals or be able to capture the excitation energy as this migrates into the hydrate sheaths (in other words, the radioprotector must fit well into the structure of water). During the formation of radicals from the biocompounds and the reactions of these, including reactions associated with the breakdown of hydroperoxide compounds, the radioprotector must effectively inhibit the free radical processes (HIn) [119−124], thus eliminating lesions of the type

$$R + HIn = RH + In. \tag{1.8}$$

The data discussed above lead to the following conclusions. Firstly, the reactions occurring in solutions of cell components corresponding to monomer unit concentrations from 10^{-4} to 10^{-2} M can involve all the ionized products of water radiolysis as well as a proportion of the excited products. Secondly, in the presence of several solutes the greatest extent of break- down may be expected from the cell components containing sulfhydryl groups, if the solution contains O_2. Here the triggering process is the formation of radicals from the solute with the participation of OH and O_2 radicals.

Knowledge of the successive stages of radiolytic transformation will enable the adoption of a suitable protection strategy. In our view, one of the best methods for elucidating the mechanism of radiolysis of biopolymers in aqueous solutions is low-temperature radiolysis in combination with various procedures for the detection of the products obtained.

MECHANISM OF THE FORMATION AND TRANSFORMATION OF BIOPOLYMERS AS DETERMINED BY THE LOW-TEMPERATURE RADIOLYSIS METHOD

Specific features of the low-temperature radiolysis method

The specificity of this method stems from the conditions of preparation (freezing of the solution), irradiation (radiolysis at low temperatures), and thawing of the samples obtained to observe the reactions of the cold-stabilized radiolytic products and subsequently analyze the products chemically. In other words, this method is distinguished by the phase state of the system examined and the structural changes occurring in this.

The samples are frozen prior to irradiation. Generally speaking, the freezing of a solution may lead to partial crystallizing-out of the solvent (ice) and of the solute, and, finally, the separation of a solid solution as a separate phase. Clearly, the radiochemical events in each of these phases must be considered in order to understand the radiolytic behavior of the system as a whole. Difficulties arise above all as a result of the different polarity of the medium in the various phases and the different conditions under which the primary physical processes of radiolysis take place in each phase of the system. This is manifested in the formation, migration and stabilization of charges and excitation in the different phases of the system, i. e., in the conditions of formation and stabilization of charges either separately or in pairs, the distances traversed by the charges or excitations along canals provided by the crystalline lattice or amorphous medium, etc.

These differences in the course of the primary physical processes in the various phases of the system are finally reflected in the nature and quantity of lesions of the tested solute. In the radiolysis of water and liquid aqueous solutions, for example, the yield of ions as determined by radiochemical kinetical procedures is about 3 per 100 eV of absorbed energy. On the other hand, measurements of the optical absorption of stabilized electrons show that the radiochemical yield of ions (e_{st}) in crystalline ice at 77°K does not exceed a few thousandths [125]. The yield of OH and H radicals in the liquid phase ranges from 4 to 12, depending on the conditions of radiolysis [68–73]; in ice at 4.2°K, the lowest temperature tested, it is approximately 1 [126, 127]. A similar picture can be expected with any compound whose phase state (or more precisely, the structure) changes with temperature.

Thawing of the solutions (from low temperatures to the melting point) causes structural changes and restores the motility of the molecules; at this point both charges and radicals escape from their traps and recombine. The entry of stabilized radiolytic products into the reactions by stages can be observed under such conditions. The radiolysis of frozen aqueous solutions at temperatures above 77°K does not necessarily represent a mere sum total of the events accompanying irradiation at 77°K and the thawing effects, as is the case, for example, with the accumulation of H and OH radicals in irradiated ice at 4.2° and 77°K [126, 127]. In the radiolysis of aqueous solutions at temperatures above 77°K it is first of all necessary to determine whether the transition from 77°K to the temperature of radiolysis alters the phase composition of the system. Such a change affects not only the nature and number of the traps stabilizing the radiolytic products but also the conditions of interaction of the active centers generated by the irradiation and during the reaction. A change in the stabilization of the active centers may influence the distribution of the radiolytic products formed by processes occurring along the track in the case of overlapping tracks). Such phenomena may alter the composition and yield of the final radiolytic products in comparison with the pattern of radiolysis at very low temperatures or in a liquid phase.

All these considerations lead to the following conclusions.

1. Vitrified solutions must be used in order to arrive at a quantitative comparison of the results obtained from the low-temperature radiolysis method and those obtained by studying liquid solutions and cells at low temperatures.

2. The low-temperature approach enables observation of the successive stages in the transformation of the radiolytic products and thus provides information on the mechanism of the process.

Detection of the primary breakdown products of the system and the primary radiochemical processes

One of the advantages of the low-temperature method is that it enables detection of the normally short-lived primary and radical-type products of water radiolysis. The earliest studies on the radiolysis of frozen aqueous solutions of inorganic acids, salts and alkalies using the EPR (electron paramagnetic resonance) method revealed the presence of atomic hydrogen, OH (O^- in alkaline solutions) and HO_2 (or O_2^-) radicals, and stabilized electrons on the basis of their optical properties [128].

These radiolytic products arise from the transformation of the ionized and excited water molecules generated by the irradiation, according to the following reactions:

$$H_2O \xrightarrow{\sim\sim} H_2O^+ + e; \qquad (1.9)$$

$$H_2O \longrightarrow H_2O^*; \qquad (1.9a)$$

$$e \rightarrow e_{st}; \qquad (1.10)$$

$$H_2O^+ + H_2O = H_3O^+ + OH; \qquad (1.11)$$

$$OH + OH^- = O^- \cdot H_2O \text{ (alkali medium)}; \qquad (1.12)$$

$$H(e) + O_2 = HO_2 \ (O_2^-); \qquad (1.13)$$

$$H_2O^* \equiv H\ldots OH \xrightarrow{OH^-} H + O^- \cdot H_2O \text{ (alkali medium)}; \qquad (1.14)$$

$$e + H_3O^+ = H_2O + H \text{ (acid medium)}; \qquad (1.15)$$

$$e + HAn = H + An^- \text{ (An } - \text{anion)}. \qquad (1.16)$$

Reaction (1.16) is of particular importance in concentrated solutions of acids and acid salts.

The above scheme of the formation and conversion of the primary products of water radiolysis agrees with more recent data [104]. It is noteworthy that the breakdown of excited water molecules into radicals $[G\ (H_2O^*) = 0.8]$ requires the previous accumulation of the radiolytic products of the system — namely, e_{st}, e_2^{2-}, O^- and O^{2-} charges — in the matrix. Similar conditions for the breakdown of excited molecules apparently arise during the radiolysis of comparatively concentrated solutions of compounds existing in the form of charged particles — namely, ions.

It has been stated that the low-temperature radiolysis of solutions of strong acids (H_2SO_4, H_3PO_4, $HClO_4$) can cause a stabilization of H_2O^+ ions in the hydrate envelopes of the acid anions [129, 130). This view, however, is open to doubt in view of more recent findings [131].

Studies carried out in the early 1950s [132–134] and during the last few years on the radiothermoluminescence (RTL) of ice and aqueous solutions have confirmed the existence of charge stabilization both in polycrystalline samples (ice) and in vitreous solutions [135, 136]. Only one of the charged particles, e_{st}, can be detected at 77°K (by the optical absorption spectrum [125]). If reaction (1.10) takes place under the conditions of radiolysis used, the ion bearing the opposite charge (barring admixtures, of course) may be the hydroxonium ion; recombination with this gives rise to the luminescence recorded by the RTL method. If both e_{st} [137] and luminescence [134] are detected in irradiated frozen solutions the ions carrying the opposite charge may be located either on the solute molecules (especially in concentrated solutions) or on the water molecules. For this reason the possible role of positively charged centers in the formation of radiolytic products and in the transformation of solutes must be taken into account when drawing up schemes explaining the mechanism of radiolysis, especially in highly concentrated solutions.

Conclusions as to the role of a particular radiolytic component of water in the radiolytic transformation of a solute may be based first of all on the comparison of the kinetics of its formation (or its radiochemical yield) in matrix solutions free from the tested compound in particular biopolymers and in solutions containing it. Alternatively, the role of the component may be assessed according to whether the presence of solutes affects the course of track processes and thus influences the formation of the primary

products of water radiolysis. Information on the role of a particular component of water radiolysis in reactions with solutes can also be obtained from
experiments on the kinetic behavior of the stabilized particles when the
sample is subjected to higher temperatures or illumination.

Let us consider the results of these studies. It must be borne in mind
that the recorded radiolytic products of water in frozen solutions, e_{st} and
H, show a marked tendency toward saturation with respect to super-high
frequency (SHF) power [138–140] and that their yield, judging from the
kinetics of their accumulation in vitreous solutions, can only be determined
at comparatively small doses (the dose-effect curves for the accumulation
of paramagnetic particles are practically linear up to 50 krad) [104]. Accordingly, a quantitative assessment can only be made on the basis of
studies in which these factors were taken into consideration.

It was found that the yield of radiolytic products of water obtained by the
irradiation of aqueous solutions of biopolymers (DNA, DNP, starch and
protein) at 77°K differs from that determined in pure matrices, the discrepancy increasing with solute concentration [26, 141–145]. At 77°K the
biopolymers examined behave differently with respect to the components of
water radiolysis. Thus, DNA and protein (human serum albumin, calf
thymus histone) in aqueous solution at 77°K lower the yield of e_{st} and H (in
vitrified matrices), whereas the yield of the oxidative component is only
slightly changed in solutions of DNA, DNP (calf thymus) and protein. In
solutions of polysaccharides or their low molecular weight analogs [143–
145], on the other hand, the yield of OH radicals declines markedly with a
rise in concentration, whereas that of e_{st} tends to increase. The yield of
atomic hydrogen in the solutions tested in an acidic or an alkaline matrix
is lower than in the matrices themselves (and is sometimes equal to zero).
Tables 1.3 and 1.4 present data on the radiolysis of solutions of DNA and
low molecular weight analogs of polysaccharides (glucose and lactose).*

TABLE 1.3. Radiochemical yield of radicals in DNA and glucose solutions irradiated at 77° K

Matrix	Radical	Concentration								
		DNA,% by weight					Glucose, M			
		0	0.3	1	5	10	0.1	0.3	1	2.5
10M KOH	e_{st}^*	3.4	3.4	3.1	2.2	0.8	3.6	3.4	4.2	4.4
	O^-*	3.4	3.0	3.0	3.0	3.0	2.5	1.8	–	0.7
	H**	⩾0.1	–	–	–	0	~0.1	~0.1	0.03	0.01
	ΣR (except O^-)	3.4	3.6	3.6	3.6	3.6	3.6	3.6	5.7	7.0
4.5 M H₃PO₄	H	0.8	0.6	0.4	0.15	0.15	0.7	0.4	0.2	0.1
	Radicals of the solute	0	0.8	1.0	2.2	2.4	–	0.3	1.0	1.6
	ΣR (except H)	1.9	2.7	3.1	4.1	4.3	2.7	2.7	2.7	3.6

* G was determined from the form coefficient.
** The G (H) values in alkaline solutions are only estimates, calculated from the change in the level of the
stationary concentration of atomic hydrogen in DNA and glucose solutions at doses greater than 0.6 Mrad.

---

* The data were obtained by us in collaboration with A.I.Pristupa and I.N.Prikhod'ko.

TABLE 1.4. Concentration of radicals in frozen aqueous alkaline solutions of lactose (10 M KOH), irradiated with γ quanta of Co^{60} at a dose of 5 Mrad at 77°K

Concen- tration of lactose, M	Number of radicals per 1 g of sample		
	total, 10^{18}	H, 10^{16}	O^-, 10^{18}
0.126	1.4	5.5	1.1
0.35	1.5	2.0	0.7
0.7	4.7 (2.8)*	1.1	0.4 (0.25)

* The concentration of OH radicals in neutral solutions is given in parentheses.

On being heated, the vitreous neutral or acidic aqueous solutions of polysaccharides and their low molecular weight analogs (mono- and disaccharides) irradiated at 77°K show an antibatic change in the concentration of radicals from the matrix (oxidative component) and radicals from the solute, i. e., the rise in temperature causes a decrease in the concentration of the former and an increase in that of the latter [146—148].

However, not all OH radicals (or $H_3PO_4 + H_2O$) turn into radicals of the solute. This depends on the structure of the solution, the concentration of the reagents, and the temperature at which the experiment is carried out. It may be assumed that reaction (1.7) takes place in such systems, or in the case of charge stabilization in an acidic medium, the reactions:

$$RH + HAn^+ = RH^+ + HAn; \qquad (1.17)$$
$$RH^+ + H_2O = R + H_3O^+. \qquad (1.18)$$

Here, some of the OH radicals apparently recombine to form H_2O_2; the yield of hydrogen peroxide depends on the temperature of radiolysis [146].

When neutral solutions of DNA and proteins are heated, the OH radicals practically do not react with the biopolymer molecule but merely recombine among themselves or with radicals of the biopolymer. The same applies to solutions of DNA and proteins containing phosphoric acid [149]. However, the fact that OH (or H_2O^+ [129]) radicals stabilized in neutral or phosphoric-acid matrices do not react with the dissolved biopolymers (DNA, proteins and DNP), on heating of the sample, does not contradict the conclusion that they may participate in reactions directly during irradiation and in the process of their formation in the hydrate sheaths (during the breakdown of excited water molecules into radicals).

That such processes can take place at 77°K is evident from three factors. Firstly, on the basis of data on the radiolysis of solutions of polysaccharides and their low molecular weight analogs, the yield of radicals from the solute increases at the expense of the OH radicals although the latter are immobile at such temperatures.*

Secondly, the exponential nature of the accumulation of radicals in the solution, including OH radicals $[C = C_\infty (1 - e)^{-kD}$, where C_∞ is the limiting

* Information regarding the possible occurrence of reaction (1.7) in frozen aqueous solutions was obtained in studies of the kinetics of the formation of radicals in solutions of inorganic salts in the presence of acceptors of OH radicals — ethanol and the triatomic alcohol glycerol /150, 151/.

concentration of the recorded radicals and k their extinction coefficient during the process of irradiation] depends on the structure of the solution and on the conditions of irradiation, and this proves that reactions involving OH radicals can take place at 77°K. Thirdly, the yield of radicals from the dissolved biopolymer (DNA, protein, DNP or polysaccharide) rises with the concentration of the solute. It may be assumed that the OH radicals detecte in frozen aqueous solutions at 77°K are those which have not reacted with the solute, and that the very fact of their detection means that they were formed and stabilized at a considerable distance from the biopolymer molecules, i. e., they originated not in the hydrate sheaths of the biopolymers but at some structural flaw far from the biopolymer molecule, possibly even in another phase, in the ice in the case of a polycrystalline sample, and not in the solution. It is for this reason that on defrosting they mostly recombine instead of reacting with the biopolymer molecules. The OH radicals in such samples are conceivably stabilized in the matrix at some distance from the biopolymer groups reacting with them, this distance being determined by the ratio of the coefficients $\dfrac{k_{\text{biopolymer}+OH}}{k_{\text{tr}+OH} + k_{\text{biopolymer}+OH}}$ which are proportional to the reaction rates during irradiation. Here $k_{\text{tr}+OH}$ refers to any trap stabilizing OH; $k_{\text{biopolymer}+OH}$ is the effective rate constant of all reactions involving the consumption of OH radicals in reactions with the biopolymer.

The absence (or reduced yield) of atomic hydrogen in concentrated solutions of biopolymers and their low molecular weight analogs may be attributed [141] to the fact that H atoms retain considerable motility even when the irradiation is carried out at 77°K; they can participate in reactions such as (1.5) or bind at double bonds:

$$\text{>C=C<}^R \quad +H= \text{>}\dot{C}-\overset{\overset{\textstyle H}{|}}{C}<^R \tag{1.19}$$

(it is known that H atoms can migrate in ice at 15°K [152]). Direct evidence for the participation of H atoms in reactions (1.5) and (1.19) was obtained in the radiolysis of frozen vitrified solutions of isopropanol and thymidine. It was found [153] that 60% of the H atoms stabilized at 77°K in a 6 M sulfuri acid solution of thymidine (0.02 M) may become attached at double bonds in the course of thawing.

When frozen aqueous solutions are radiolyzed at higher temperatures, at which the oxidative component of radiolysis (OH radicals) is motile, the yield of solute breakdown is greater than at 77°K in the same system.

As noted above, the participation of the radiolytic products of water in reactions with the solute causing its breakdown, during low-temperature radiolysis, depends largely on the structure and phase composition of the system. This is illustrated in the results of the radiolysis of aqueous solutions of biopolymers and their low molecular weight analogs irradiated at different temperatures (Table 1.5) [54, 134–140, 154, 156]. The $G(R)$ values obtained using an EPR-2IKhF spectrophotometer at doses of 0.7– 1 Mrad can be regarded as effective. These findings clearly demonstrate the relative effect of the phase composition of the system on the degree of radiolytic breakdown of the solute.   It is evident from Table 1.5, for example, that the yield in polycrystalline samples changes (at first rises, then declines) with the transition to temperature ranges delimited by points at which there is a change in the ice structure, i.e., between 77° and 120°K (increased yield) and between 120° and 160°K (decreased yield).   It follows that the results obtained reflect changes in the conditions under which the radicals appear, become stabilized, and interact in the system.

TABLE 1.5.   Yield of radicals in frozen aqueous solutions

Irradiated system	T, °K	G (R)	Irradiated system	T, °K	G (R)
DNA (30%)	77	0.7	2-Deoxyribose (0.75 m)	77	0.6
	118	2.3		123	2.2
DNA (10%)	77	0.5		153	1.0
	118	1.0		198	0.5
	153	0.3	2-Deoxyribose (1.5 M)	123	1.6
	198	0.1	Arabinose (0.75 M)	77	0.4
DNA (5%)	77	0.5	Arabinose (1.5 M)	77	0.6
	153	0.15	Gentiobiose (0.3 M)	77	0.4
	198	0.4		123	0.5
DNA (1%)	153	0.1		203	0.4
DNA* (20%, gel)	123	0.6	Cellobiose (0.3 M)	77	0.4
DNA* (2.5%)	77	0.3		123	0.5
DNA* (2%)	153	0.6		203	0.7
DNA* (0.6%)	77	0.3	Lactose (0.3 M)	123	0.7
DNA* (0.2%)	123	0.4	Lactose (0.03 M)	123	0.1
RNA (saturated solution)	153	0.7	Maltose (0.3 M)	77	0.7
Starch:				123	0.4
waxy (5%)	77	1.9	α-Methyl-D-glucoside	77	0.3
maize (5%)	77	2.2	α-Methyl-D-galactoside	77	0.4
amylose (5%)	77	2.6	β-Methyl-L-arabinoside	77	0.7
Glucose (0.56 M)	77	0.4	Glycyl-glycine (1 M)	123	1
	163	0.6	Glycyl-valine (0.25 M)	123	2
	198	0.4	Glycyl-leucine (0.25 M)	123	1.2
2-Deoxyglucose (0.6 M)	77	0.5	α-Alanine-α-alanine (1 M)	125	1
Ribose (0.67 M)	77	0.5	β-Alanine-β-alanine (1 M)	125	1.4
	123	1.2	Glycine (1 M)	125	1
	153	1.2	α-Alanine (1 M)	125	1.7
	198	1	β-Alanine (1 M)	125	1.4
Ribose (1.34 M)	123	2.5			

* Molecular weight of DNA — $6 \cdot 10^6$ daltons.

Analysis of the available data on the low-temperature radiolysis of aqueous solutions of biopolymers and their low molecular weight analogs shows that at low temperatures (starting from 77°K) the radiolytic products of water (or, more precisely, the oxidative and reductive components of the radiolysis) enter into reactions with the solutes.   The maximum overall yield of radicals in such systems is obtained in vitreous solutions and corresponds to the breakdown of slightly more than 4 molecules of water. If the solutes (biopolymers and their low molecular weight analogs) function as acceptors predominantly with respect to one of the components of water radiolysis and prevent its recombination in the cell, they will also bring about a stabilization of the second component of the radiolysis of water. For this reason the observed maximum overall yield of radicals is 7−8 within the limits of accuracy of the experiment (see Table 1.3). A similar value of G was obtained in the radiolysis of highly concentrated solutions of carbohydrates (about 90%) [157].

Since radicals such as OH obviously cannot diffuse at 77°K in the vitreous matrices studied, it may be assumed that the dissolved compounds (OH⁻, acid anion and biological material) bring about an actual breakdown of primary active formations of the H_2O^+ type from water into ions or radicals; in other words, the triggering mechanisms for the transformation of the solutes in these cases lie in the radiochemistry of the water molecules contained in the hydrate sheaths of these solutes. This means that the breakdown into ions and radicals and the reactions of these particles with the solutes depend on the structure of the matrix and the reactivity of the solutes.

Another important problem to be solved by the low-temperature radiolysis of frozen aqueous solutions of biopolymers concerns the site of attack on the biopolymer molecule by the products of water radiolysis, i. e., the original location of the free valence in the macroradical. We shall discuss this topic with reference to the radiolysis of solutions of DNA, polysaccharides, and proteins.

Site of the lesion in DNA. The substrate breakdown taking place during the low-temperature radiolysis of frozen aqueous solutions of DNA involves the participation of e, H, H_2O^+ and OH, though at 77°K the reductive component of radiolysis performs the predominant role [135, 137, 141−143, 158−161]. A line caused by DNA radicals, possibly the sum of the singlet and doublet lines, appears in DNA solutions on the spectrum of matrix radicals at a g factor of ∼2. The intensity of this line is proportional to the concentration of DNA in the solution. Conclusions as to the nature of the DNA radicals may be drawn by comparing the available data on the EPR spectra of samples of DNA and its components, i. e., compounds such as the phosphate residue of 2-deoxyribose, nucleic acid bases, nucleosides and nucleotides in both aqueous solutions and dry preparations [162−165]. The pattern of the EPR spectra of solutions of phosphoric acid or its salts differs from that of DNA solutions [141, 142, 146]. Solutions of sugar-phosphate, carbohydrates and phosphate ion at equimolar concentrations or even containing phosphate ion at a concentration five times greater than that of the carbohydrates show EPR spectra of radicals originating from the carbohydrate, rather than those typical for solutions containing the phosphate ion [148].

It was shown experimentally that the formation of these radicals in frozen aqueous solutions results from reactions involving the oxidative component of water radiolysis, i. e., OH or H_2O^+ radicals [148]. It follows from these findings that even if the initial damage affects the phosphate component of the phosphate:sugar mixture (1:1) or the sugar-phosphate, the free valence is finally located in the sugar component. It appears that such systems initially generate a radical (cationic radical) by the removal of an electron from the O − P bond of the phosphate fragment when this is attacked by OH or H_2O^+ radicals (indirect mechanism), or as a result of the direct effect of the radiation.   Next, either an $O - P$ bond is cleaved [167], or the radical first isomerizes as the valence passes from the phosphate residue to the sugar component, and then the $O - P$ bond breaks apart.

The OH (or H_2O^+) radicals can also directly attack the sugar component. In this case the primary radical appears according to reaction (1.7) by an interaction between DNA and H_2O^+ — the cationic radical.

According to EPR measurements on irradiated frozen aqueous solutions of 2-deoxyribose (or other carbohydrates [137, 147]), the primary macroradical apparently bears the free valence at C'_1.   This radical may later isomerize.   In 2-deoxyribose solutions exposed to UV irradiation, for example, the initial doublet disappears and is replaced by a line with five equidistant components, while the total concentration of radicals remains unchanged [137].   The free valence of this radical apparently lies at C'_4 or C'_5.   A similar radical with a quintet hyperfine structure (HFS) was detected in irradiated monocrystals of 2-deoxyribose [168, 169].   The isomerization of the primary macroradical in DNA solutions possibly causes the transfer of an H atom from C'_2 to the nitrogen base, thus creating a thymine-type radical [137].

Judging from the radiolysis of liquid [170] and frozen aqueous solutions of phosphates substituted to varying degrees [171], the electron can participate in the detachment of a proton from hydrogen-containing compounds ($H_2PO_4^-$, HPO_4^{2-} or H_3PO_4), yielding atomic hydrogen.   The role of e_{aq} in the breakdown of DNA at the phosphate group is minor.   Experiments with model aqueous solutions of carbohydrates have shown that e_{aq} can also react with the sugar fragment of DNA, though the maximum contribution of this process in the overall breakdown at the sugar fragment (including the reactions of the OH and H radicals) does not exceed 10% [85].   Experiments with glucose solutions in an alkaline matrix [141] indicate that the stabilization of the electron by the sugar fragment prolongs its life span as an active particle.   Evidently, in the radiolysis of DNA solutions the sugar component of the DNA can stabilize the electron.   Under such conditions there is a greater probability that the electron will participate in reactions involving a change in the sugar fragment, although it does not react directly with this component but rather with other DNA constituents, such as nitrogen bases or radicals formed from DNA (secondary processes).   Generally, the electron attacks the nitrogen bases [141, 166, 172—175].

As noted previously (p. 12), the atomic hydrogen formed according to reactions (1.3) and (1.4), like OH, manifests itself during the breakdown of excited water molecules by attacking the sugar component of DNA and the bases according to reactions (1.5) and (1.18).   The latter reaction, in particular, yields a thymine-type radical.   Reactions involving the addition of H atoms at double bonds have been detected in dry preparations of bases, nucleosides, nucleotides, and nucleic acids [176—184].   It was found that the activation energy for the addition of H at double bonds of the thymine base in dry preparations of thymine is 1.2 cal/mole [183].

In connection with the directions of attack by H atoms on bases it is pertinent to note the results of EPR measurements made on an alloxanthine monocrystal bombarded with H atoms.   It was found [185] that H atoms can attach themselves at $C = O$ double bonds of the pyrimidine analog, forming a radical in which the unpaired electron reacts with one proton. This type of interaction of H radicals with bases may also occur, in principle, during the radiolysis of aqueous solutions of DNA.

That H atoms react directly with the $C_5 = C_6$ double bonds of thymine bases in frozen solutions is evident from several studies [141, 153, 172, 186]. It was found that defrosting of irradiated solutions of thymidine $(0.02\,M)$ in sulfuric acid [120] and of thymine [0.032 M] in alkali [172] yields thymine-type radicals. One of these studies [186] involved the photolytic generation (by means of a SVDSh-1000 lamp) of H radicals according to the reaction: $Fe^{2+} + H_2O + h\nu = Fe^{3+} + OH^- + H$ in solutions of $FeSO_4$, H_2SO_4 and DNA, causing the accumulation of thymine-type radicals at temperatures from 110 to 120°K — a range at which all the H atoms became mobile, entered into reactions and vanished from the "field of vision" of EPR.

Thymine-type radicals can also be generated directly during the irradiation of strongly acidic or alkaline solutions of DNA at 77°K (by the addition of H at the double bond of the thymine base), i.e., in DNA preparations that are already partly denatured and depolymerized. In this case, the H atoms evidently have free access to the double bond [141] (contrary to the situation existing in solutions of native DNA preparations in a neutral medium) or alternatively the thymine radical is formed in two rapid stages: first, addition of an electron, then of a proton. It may also be assumed that the formation of the thymine radical in these cases results from the localization of the excitation at a water molecule situated in the hydrate sheath of the thymine base; the breakdown of this molecule yields H and OH radicals which subsequently participate in the reaction.

The EPR spectrum of the irradiated sample (a neutral solution of a native DNA preparation) changes with a rise of temperature, becoming more complex. When the temperature reaches 150—180°K a thymine-like structure appears in oxygen-free solutions [137]. The formation of such a structure on heating DNA solutions has also been detected by other workers [166, 187].

Similar changes in the EPR spectrum may be observed when a γ-irradiated solution of a native DNA preparation previously warmed to 130°K is exposed to unfiltered UV light. These phenomena, associated with the phototransformation of the radicals (the doublet HFS turns into a thymine-like structure), are observed in vitreous solutions of DNA either acidified with phosphoric acid, alkaline or neutral (in the presence of glucose). According to the evaluations made, the yield of thymine-type radicals by thermal annealing or by photo-annealing (by unfiltered UV light) does not exceed 10—20% of the original number of radicals formed from the DNA, mainly radicals with a doublet HFS.

It follows from these experimental results that thymine-type radicals may be generated by two mechanisms: 1) by addition of an electron to the thymine base to form a negatively charged paramagnetic particle which then unites with a proton (first pathway); 2) by the formation of a radical of which the free valence lies initially not on the thymine base but on another fragment of the biopolymer, for example, the sugar and the subsequent isomerization of the radical so that the free valence is situated at C_5 of the thymine base (second pathway). Experiments with DNA solutions in light and heavy water [166] have shown that the proton can migrate from the complementary base to the damaged thymine which has previously accepted an electron at the $C_5 = C_6$ bond, forming the radical $\diagup \!\!\!\! \overset{.}{C} \!\!-\!\! C \!\!\diagdown \!\!-$.

As shown by the low-temperature radiolysis of nitrogen bases, nucleosides and nucleotides [172–175, 186, 188], the protonation of anionic radicals M^- in alkaline and acidic vitrified matrices may take place (heating samples previously irradiated at 77°K) as a result of their interaction with water molecules or directly with H^+ (H_3O^+). It was found that the anionic radicals formed by the addition of e to pyrimidine-type nitrogen bases have a doublet HFS ($\Delta H_{max} \approx 22-28$ gauss) and in the case of purine bases — singlet EPR lines ($\Delta H_{max} \approx 15$ gauss) [188]. The protonated radicals MH have a more complex HFS. The anionic radical formed during the defrosting of irradiated solutions of thymine reacts with the proton and turns completely into a thymine-type radical.

The observed consecutive change in the EPR spectrum of an irradiated DNA solution with the formation of a thymine-type radical can also be explained in terms of an isomerization or a temperature- or UV-induced conversion into a radical in which the free valence is situated in the sugar component of DNA. In this case the doublet spectrum observed ($\Delta H_{Makc} = 29$ gauss) should be attributed to the 2-deoxyribose radical.

Comparison of the radiochemical properties of frozen solutions of DNA and monosaccharides, including ribose, 2-deoxyribose and glucose, reveals two important features. Firstly, the yield of atomic hydrogen stabilized at 77°K decreases with a rise in the concentration of the compound examined; the yield of the oxidative component of radiolysis similarly decreases — immediately at 77°K in the case of sugars, and in DNA at higher temperatures with a simultaneous rise in the yield of doublet-type and thymine radicals from DNA (at higher temperatures of radiolysis, as noted above) [137, 141]. Secondly, solutions of DNA and carbohydrates contain radicals with the same spectral characteristics (practically identical doublet HFS lines with $\Delta H_{max} \approx 29$ gauss) and relaxation characteristics whatever the nature of the matrix (highly alkaline, highly acidic, neutral), judging from the capacity for saturation with the rise of HSF power [141]. In other words, the free valence "feels" the immediate surroundings of the radical but not the influence of the matrix.

Analysis of the published data on the low-temperature radiolysis of aqueous solutions and dry preparations leads to the conclusion that the free valence in the sugar component of DNA lies initially at C'_1 (provided that the two hydrogen atoms at C'_2 are not equivalent as regards their reaction with the unpaired electron), or possibly at C'_5 [189]. A radical may be formed by the removal of H from C'_1 since this atom is the most mobile and therefore more likely to be detached than other atoms as a result of its position in the molecule (like C_α with respect to the oxygen atom of the 2-deoxyribose ring component and the aromatic component of DNA, i.e., the base).

It may be assumed that UV light or temperature causes the transfer of one of the H atoms from C'_2 to the $C_5 = C_6$ double bond of the thymine base, leading to isomerization of the macroradical. During the photolytic or thermal annealing of DNA radicals, only $10-20\%$ of them turn into radicals of the thymine type, while the overall yield of DNA breakdown remains practically unchanged. This suggests that such isomerization processes, involving the transfer of hydrogen from the sugar component to the base, occur not only in thymine but also in other nucleosides containing a damaged sugar component [189, 191]. However, the radicals formed by the addition of H at the double bonds of bases are possibly less stable than the thymine radical or more difficult to detect owing to their narrower spectra (against the background of lines produced by the thymine and other radicals Indirect evidence for the occurrence of such processes, involving a migration of hydrogen from the sugar component to the base in nucleosides, can be found in mass spectrophotometric measurements [192]. It was found [192] that the main trend of the breakdown of the primary positive ion generated in the mass spectrophotometer is toward the destruction of the sugar component of the nucleoside and cleavage of the $C'_1 - N$ bond, which releases the base with or without the addition of another H atom (or a protor — it is impossible to unequivocally decide this point since the particle recorded is the positively charged ion of the base) from the $H_2C'_2$ group of the sugar. The yield of the latter process (involving a regrouping of atoms, as a result of which the H atom or the proton settle on the base) depends on the nature of the sugar component. Purine nucleosides show a high yield of base liberation without the introduction of an additional H.

If the free valence of the initially formed DNA macroradical is located at C_5, it is more complicated to explain the observed change in the EPR spectra during the thermal and photolytic treatment of the sample, but never theless possible. Thus, the radical formed as a result of attack on the 2-deoxyribose unit may be produced by reactions (1.5) and (1.7) or with the participation of H_2O^+ ions in concentrated solutions, as noted previously:

$$
\begin{array}{lllll}
H_2O^+ & Ph & Su & H_2O & Ph^+ & Su, \\
Ph^+ & Su & Ph & Su^+; & & \\
Su^+ & H_2O = R \cdot + H_3O^+, & & &
\end{array}
$$

$$(1.20)$$

$$(1.21)$$

where Ph and Su are respectively the phosphate and sugar fragments of DNA. This scheme explains the data obtained from EPR measurements of frozen aqueous solutions of sugars and DNA. Further evidence for the occurrence of these processes has been obtained from low-temperature radiolysis of dry preparations and analysis of the interaction of a halogen-substituted nucleoside — uridine — with H atoms generated by a silent gaseous discharge [190, 191].

EPR studies of the radiochemical properties of nucleic acids, nucleotides nucleosides, 2-deoxyribose, ribose and nitrogen bases (dry preparations) have shown that the yield of radicals in nucleoside and nucleotide preparations is greater than that obtained in the radiolysis of pure bases [142, 162].

The EPR spectra of nucleotides and nucleosides at temperatures higher than $77°K$ and at comparatively large irradiation doses corresponded to the typical spectra of the pure bases. In a study of the radiolysis of dry polycrystalline samples of nitrogen bases, nucleosides and nucleotides at $130°K$

performed in our laboratory (in collaboration with Yanova) we were able to detect only the EPR spectra of radicals originating from the damaged bases of the irradiated nucleosides and nucleotides and not those of radicals arising from the sugar fragment; similar results have been reported by Müller et al. [162]. Thus, the radiolysis of uridine, uridylic acid, thymidine and thymidylic acid yields EPR spectra typical of nitrogen base radicals arising as a result of the addition of an H atom at the $C_5 = C_6$ double bond. To explain the observed effects in the nucleosides and nucleotides it was assumed that the original lesion appears in the sugar fragment and that the radical then isomerizes immediately during irradiation even at these temperatures, so that one of the H atoms of the sugar fragment migrates to the double bond of the nitrogen base.* A radical with a similar HFS can conceivably appear also as a result of a direct interaction of H atoms with the nitrogen bases. Experiments with monocrystals or uridine derivatives bombarded with H atoms [189–191] have shown that these compounds (nucleosides) actually yield two types of radicals — one formed by the detachment of an H atom from the sugar component (from $C'_5 - H$), the other by the addition of H at the double bond of the nitrogen base.

In view of the published evidence [141], the phenomena detected by the EPR method can also be explained by a third scheme which actually represents a modification of the first. The electrons generated during the irradiation of DNA solutions are stabilized by the sugar fragment and are then captured by the (four) nitrogen bases of the DNA, yielding anionic radicals. On photolysis or thermal annealing of the sample, the anionic radicals acquire a proton (the anionic radical reacts with H_2O) and turn into neutral paramagnetic particles; pyrimidines thus yield a radical formed by the addition of H at the $C_5 = C_6$ double bond, and purines — less stable radicals which may subsequently break apart at the $C - C$ and $C - N$ bonds.

These hypotheses can be tested directly by obtaining anionic radicals of the nitrogen bases directly in a liquid phase, electrochemically for example, then recording and analyzing their EPR spectra. The possibility that the electron participates in the formation of radicals as a result of electron capture by the bases cannot be ruled out since the illumination of irradiated dry preparations of DNA [192] bleaches them, destroys about 90% of the paramagnetic centers and alters the course of some macromolecular chemical processes. All this illustrates the important role of stabilized charges in the overall effect of radiolysis of DNA.

To sum up, the available data on the low-temperature radiolysis of DNA solutions indicate that the primary events leading to DNA lesions involve the following processes.

1. Detachment of H radicals from the sugar fragment by H atoms at 77°K and by OH radicals at higher temperatures.

2. Interaction of the oxidative component of radiolysis and DNA via the phosphate group to give a radical located in the sugar fragment, apparently possessing the same structure as in the former case [initially a cationic radical is formed according to reactions (1.20)].

3. Addition of the H radical at the double bond in denatured DNA preparations at 77°K.

---

* The effects reported in [126] were later interpreted in a similar manner [163].

4. Addition of OH radicals at double bonds of the DNA bases (according to experiments using the flow technique) [194].

5. Capture of e_{aq} by the nitrogen bases.

The sequence of transformations of the initially formed radicals and their disappearance with a rise in temperature of the solution can be outlined as follows.

1. Isomerization of the macroradical containing the damaged sugar fragment and migration of the H atom to the double bond of the nitrogen bases.

2. Transformation of the anionic radical by an interaction with water or by the addition of a proton H^+ from the complementary base (or recombination of opposite charges).

A further rise in temperature causes the disappearance of the stabilized OH radicals and of some of the radicals formed from DNA. The OH radicals disappear faster than those originating from DNA [149]. The disappearance of the radicals is associated with the reactions (1.3) and

$$R_{DNA} + OH = \text{products.} \tag{1.22}$$

The formation of hydrogen peroxide in frozen DNA solutions was detected by chemical methods after defrosting of the solution. Chemical procedures have also revealed the presence of glycols formed during the radiolysis of liquid aqueous solutions [58] according to reactions such as (1.22).

If the radiolysis of DNA takes place in solutions containing additives such as molecular oxygen, the sequence of radical transformation includes certain other reactions judging by the following experimental data.

Experiments with high molecular weight native DNA preparations have shown that a rise in temperature in the presence of dissolved molecular oxygen yields an assymetrical spectrum ascribed to a peroxide-type radical RO_2 while no thymine radical can be detected. This suggests the occurrence of processes of the type

$$R + O_2 = RO_2. \tag{1.23}$$

A further rise in temperature causes the disappearance of this radical [137], apparently through the formation of hydroperoxide compounds which are recorded among the radiolytic products after defrosting of the solutions.

If frozen solutions of DNA are radiolyzed in the presence of other additives acting as radioprotectors, in the form of sulfhydryl compounds ([166, 158] and our own data) or p-benzoquinone and propyl gallate (PG) [158, 195, 196], the radicals detected on defrosting are not of the thymine type but depend on the additive present in the solution, as is the case with molecular oxygen. Sulfhydryl compounds yield a radical with an asymmetrical spectrum, and bearing the unpaired electron on the sulfur atom (paramagnetic particle of the organic sulfur radical type [196] or anionic radicals [198, 199]); benzoquinone (an efficient electron acceptor) produces a narrow singlet line corresponding to the anionic radical of this compound and the propyl gallate radical (the structure of the spectrum is more or less the same as in the former case). Here the total concentration of radicals in the sample remains practically constant, and the curves showing the change in the concentration of radicals from DNA (declining) and from PG (ascending) are antibatic.

It is evident from these observations that the transformation of radicals in this system involves migration of the free valence between the DNA macroradical and PG at a ratio of approximately 1:1. According to radiothermoluminescence (RTL) data, the transfer of the charge, obviously an unpaired electron, is most effective in a frozen solution containing one PG molecule per DNA nucleotide. RTL observations show that defrosting of the irradiated samples in these conditions (within the range of temperatures in which there is a transition from DNA radicals to PG radicals) lowers the luminescence peaks in solutions containing DNA and PG.

If the luminescence results from the recombination of charges and these charges, as it seems, are located in the DNA structure, it may be assumed that when PG is introduced into the solution it combines with the DNA molecule to form a complex within which the charge or uncompensated spin [135] can migrate from the damaged DNA molecule to PG. The subsequent recombination of this charge with an opposite charge takes place on the PG component, which functions as an extinguisher. In other words, we interpret the protective action of PG as a capture of the charge (electron) initially located in the DNA.

The nature and properties of the radicals formed in aqueous solutions of proteins, protein-like compounds and DNP. Experiments on the radiochemical properties of the protein-water system have shown that the formation of radicals from dissolved biopolymers [143, 154] and their low molecular weight analogs — peptides [152, 200, 201] and also from amino acids [54, 200–204] in frozen aqueous solutions at the temperature of liquid nitrogen and higher temperatures results from the indirect and direct effects of radiation on the system. In the case of the indirect action of irradiation, the solute reacts at 77°K mainly with the reductive component of water radiolysis, i.e., the H and e_{aq} radicals. The latter, detected as e_{st} by a singlet EPR line in aqueous solutions of proteins, is stabilized at a low yield and is bleached by visible light [143, 154]. The yield of OH radicals detected by the EPR method in the 77–125°K temperature range is comparable to that of the solute. A rise in the temperature of radiolysis of the frozen aqueous solutions within this range increases the yield of radicals from the solute owing to the reduced yield of OH radicals [54, 152, 201].

Several authors [54, 152, 200, 201] have pointed out that the EPR spectra of the radicals of dissolved compounds of this type are essentially the same as the EPR spectra recorded in the radiolysis of dry preparations. This means that radicals of identical structure may be formed in the systems discussed.

The EPR spectra of radicals formed from dissolved proteins (human serum albumin) [143–145] and protein-type compounds such as calf thymus histone and Thiogel [187, 205] result from the superimposition of several EPR lines: a doublet line, one or several singlet lines, and multicomponent lines with a total width of about 90 gauss in histone solutions and up to 120 gauss in Thiogel solutions.

Comparison of the EPR spectra of irradiated frozen solutions of proteins and their low molecular weight analogs with those of the radicals of the same compounds, recorded at low temperatures in dry preparations bombarded with H atoms generated by a gaseous discharge [206, 207] or those of various polypeptides [208, 209], peptides [152, 210—217], amino acids and proteins [218—225] irradiated with accelerated electrons or γ-quanta of Co^{60} gives rise to the conclusion that the EPR spectra of protein radicals in frozen aqueous solutions result from the superimposition of the lines corresponding to the primary radicals formed by the following reactions.

1. Addition of H and OH to aromatic amino acids (formation of a singlet or triplet with a total line width of about 40 gauss [206, 210, 213]).

2. Detachment of H atoms according to reactions (1.5) and (1.7) with the formation of a free valence in the biopolymer chain [207—209], leading to radicals of the type

$$\cdots -\overset{\overset{\textstyle O}{\|}}{C}-N-\overset{\overset{\textstyle \cdot}{}}{\underset{\underset{\textstyle R}{|}}{C}}-\overset{\overset{\textstyle O}{\|}}{C}-\underset{\underset{\textstyle H}{|}}{N}- \cdots$$

$$\underset{\textstyle H}{|}$$

3. Detachment of an H atom from side chain amino acid residues including aliphatic [152, 210]* and sulfur-containing [199, 210] residues; this process yields respectively wide lines (up to 120 gauss) and an asymmetrical spectrum (if the free valence is located on the sulfur atom).

4. The interaction of the electron with the substrate brings the free valence to the oxygen atom of the peptide bond (giving a doublet HFS with an interspace of about 20 gauss, caused by the H atom participating in the hydrogen bond $N-H...O=C$ [219]).

5. Reaction of the electron with the $C=O$ double bond with the subsequent liberation of ammonia from the terminal group and the formation of a radical with the free valence located at C_α with respect to the carbonyl group according to the mechanism described by Willix et al. [226], or localization of the electron on the aromatic rings and on the sulfur atom, leading to anionic radicals [223—225].

The site of attack by the products of water radiolysis and the location of the free valence in the damaged protein molecules and in the molecules of their precursors depend largely on the dissociation state of the molecules [198].

Experiments on the thermal annealing of radicals in frozen aqueous solutions of proteins and their low molecular weight analogs have shown that with a rise in the temperature of the sample the molecular mobility of the different parts of the system gradually reappears; the radicals stabilized in the system consecutively either vanish or turn into radicals of a different structure. It was found [54, 143, 154, 191, 202], for example, that the radicals originating from water — namely, OH and e_{st}, disappear in frozen aqueous solutions of proteins [143, 154, 187, 205] at a temperature of 120—130°K. In aqueous solutions of sulfur-containing amino acids, peptides (glutathione) or other organic sulfur compounds (penicillamine), from 5 to 30% of the OH radicals disappear with a rise in temperature and yield solute radicals with the free valence located on the sulfur atom [200].

---

* The occurrence of reactions such as (1.7), leading to the formation of radicals from the aliphatic ends of amino acids, has been demonstrated with liquid solutions of amino acids by the EPR method using the flow technique for the generation of OH radicals [211].

With a rise in temperature the radicals originating from the solute may isomerize [152] or react with intact molecules to form less reactive radicals [216] (intramolecular or intermolecular migration of the free valence).

In the presence of oxygen and radioprotectors in the protein solutions, processes leading to the formation of peroxide radicals [143, 154] or involving migration of the free valence from the damaged biopolymer to the radioprotector may take place. Thus, the defrosting of irradiated solutions of Thiogel (9%) and penicillamine (2.25%) at temperatures ranging from 160 to 220°K yields penicillamine radicals of the $RC(CH_3)_2S$ type instead of protein-type radicals [205]. It was also found that a directional migration of the free valence (or more precisely, energy) to the radioprotector may take place in such solutions: the radiochemical yield of radicals from the radioprotector in solutions of these compounds is more than 10 times greater than the yield of radicals from the radioprotector in a solution of the latter. The observed similarity of the transformation of radicals in protein-radioprotector systems in frozen solutions and dry preparations (mixtures) indicates that the radioprotective effect consists in the formation of a complex between the macromolecule and the sulfur-containing compound [205, 227, 228]. Protein radicals exist in solution at temperatures of up to 200°K [158].

The main pathways of the transformation of the protein molecule under the influence of the radiolytic products of water may be outlined as follows on the basis of the above data regarding the nature of the primary radicals and the changes in the EPR spectra detected with a rise of temperature of frozen aqueous solutions of proteins.

1. The OH radicals attack mainly the aromatic amino acid residues, causing cyclic cleavage and yielding peroxide compounds if the radiolysis takes place in the presence of molecular oxygen.

2. The radicals formed from proteins by the detachment of an H atom and localization of the free valence in the middle of the biopolymer chain can cause breaks in the chain as a result of their subsequent transformation. Moreover, the intermolecular migration of the free valence leads to the consecutive transformation of these radicals culminating in a more stable radical which then enters into the reaction to yield the final products of radiolysis.

3. The reaction of the electron with the double bond of the carboamino group may lead to deamination if the carboamino group involved occupies a terminal position on the biopolymer.

Studies of the radiolysis of frozen aqueous solutions of DNP and histone have revealed the formation of OH, e_{aq} and radicals from DNP in the frozen solutions of the latter; the yield of radicals from the dissolved polynucleoproteid is proportional to the concentration of DNP in the solution [143, 144, 154, 155]. The formation of radicals from DNP involves the participation of both the OH radical and the reductive components of water radiolysis, namely, e_{aq} and H. The total yield of radicals in DNP solutions is the same as in ice irradiated at 77°K. The OH radicals stabilized at 77°K in aqueous solutions of DNP do not react with the dissolved biopolymer on defrosting of the solution. In the case of DNP, the spectrum obtained after the disappearance of the OH radicals with defrosting of the sample consists of a poorly resolved doublet line with $\Delta H_{max} \approx 29$ gauss.

It is evident from the EPR spectra of irradiated DNP solutions that the free valence of the irradiated DNP radical lies initially in the protein component. On further defrosting of the sample, the free valence migrates to the DNA [144, 154]. The bombardment of dry preparations of DNP with discharge-generated H atoms similarly yields radicals located in the protein component [181]. These radicals are presumably identical to those formed as a result of irradiation.

Radiolysis of polysaccharides and their low molecular weight analogs. Several publications [143, 145, 229] dealing with the low-temperature radiolysis of polysaccharides in aqueous solutions at 77°K stress the crucial role of the oxidative component of water radiolysis in the formation of radicals from the dissolved biopolymer (different types of starch). The reductive component of water radiolysis — atomic hydrogen in these systems — participates in the detachment of H atoms from the solute molecule in the same way as the OH radicals. Under such conditions of radiolysis e_{aq} practically does not react with the solute but is stabilized on the sugar fragment at a comparatively high yield. The EPR line of e_{aq} shows a marked tendency for saturation with SHF power and vanishes together with the color of the sample under the influence of visible light.

The line totally disappears following exposure to visible light in transparent alkaline matrices [85]. The time required for the complete bleaching of e_{st} (disappearance of the EPR line and of the violet color) in transparent alkaline matrices increases with the concentration of the carbohydrate in the solution. Thus, total bleaching of e_{st} in glucose solutions takes several minutes at a concentration of 2.5 M in 10 M KOH, whereas in 0.01 M glucose solutions this (first-order) process requires only a fraction of a minute and in pure KOH solutions — even less under the same conditions of illumination [141]. In neutral solutions of sugars (at concentrations of up to 1−2 M) the singlet line does not disappear completely after exposure to visible light (although light does penetrate through the apparently opaque samples). All these facts indicate [19, 146, 147] that the stabilization of the electron on the carbohydrate molecule yields a molecular ion M⁻. This particular form of stabilization of the electron by the carbohydrate molecule — the formation of M⁻ — is typical of the liquid-phase radiolysis of carbohydrates and influences the composition of their transformation products [230].

The accumulation of radicals in frozen aqueous solutions of polysaccharides under the influence of radiation follows an exponential course, as is the case with other biopolymers. Solutions of different polysaccharides and their low molecular weight analogs show a great similarity with each other not only with respect to the kinetics of the accumulation and transformation of radicals during irradiation and the thermal or photolytic annealing of radicals in the irradiated samples but also in the structure of the EPR spectra of the radicals generated [19, 85, 141, 146−149, 154, 161, 231−238]. Such uniform behavior indicates that the primary processes of radiolysis are less diverse in these systems in comparison with those discussed above.

Analysis of the data on the yield of radicals indicates that the radiolytic properties of these systems depend mainly on the structure. This explains the comparatively low radiochemical yield of radicals from dissolved monosaccharides and disaccharides, and the comparatively high yield of radicals from polysaccharides (various types of starch) in solutions of comparable concentration (5% by weight, for example) at 77°K.

However, the yield of radicals from the sugar increases with the concentration of the latter in the solution even at 77°K to a maximum of 8 (in 90% solutions of fructose [157]), i. e., 4 if one takes into account that about half the paramagnetic particles must be attributed to e_{st}. A comparable yield of radicals can be obtained at lower concentrations of dissolved monosaccharide (glucose [141]) if the radiolysis is carried out in vitreous matrices in the form of strongly alkaline or strongly acidic solutions.

The high yield of radicals from polysaccharide solutions in comparison with solutions of mono- and disaccharides can be attributed to the fact that the biopolymer (starch) molecule structures the solution to a greater extent than the monosaccharide, thus approximating the conditions of radiolysis in a vitreous matrix and ultimately those in a liquid phase.

Differences in the structure of the carbohydrates and hence in their reactivity toward the radiolytic products of water may also be responsible for the observed diversity in the radiochemical yield of radicals (beyond the limits of experimental error) from dissolved polysaccharides and their low molecular weight analogs (see Table 1.5) in solutions of equal concentration.

Contrary to its behavior in the systems discussed above, the stabilized OH radical can react with the biopolymer molecule following a rise in temperature [145, 146, 147, 231]. Studies of frozen aqueous solutions of monosaccharides, disaccharides, glycosides and polysaccharides have shown that the degree of participation of the stabilized OH radicals in reactions with the solute depends on the temperature of annealing and the concentration of the carbohydrate, i. e., on the structure of the solution and that of the biopolymer. Between 10 and 30% of the stabilized OH radicals can participate in the formation of radicals from the solute: the curves showing the concentration of OH radicals and radicals from the solute are antibatic. According to a study [205] of the radiolysis of frozen 10% aqueous solutions of the dextran Sephadex-50 (a structured polymer of glucose), about 40% of the radicals stabilized at 77° recombine with a rise in temperature (the solutions contained 2.25% penicillamine).

The nature of the radicals formed during the radiolysis of polysaccharides in frozen solutions can be clarified by a comparison of EPR measurements of polysaccharides and their low molecular weight analogs — mono- and disaccharides. In studies of the radiolysis of frozen aqueous solutions of seven monosaccharides and five disaccharides, the irradiated carbohydrate solutions were found to have closely similar EPR spectra. In all cases, without exception, the EPR spectra of the polysaccharides and low molecular weight analogs at 77°K showed a doublet line with an interspace of about 20 gauss and $\Delta H_{max} \approx 30$ gauss at a g factor of 2. The doublet HFS results from the interaction of the unpaired electron with the hydrogen nucleus. Comparison of the EPR spectra of radicals from monosaccharides of different structure shows [147, 231] that the doublet radical is formed as a result of the detachment of an H atom from C_1 (according to data on the radiolysis of solutions of arabinose — the only monosaccharide where detachment of the H atom from C_1 yields unequivocally a radical in which the unpaired electron can react with one hydrogen atom). In view of the high yield of the radical with a doublet HFS in solutions of polysaccharides, mono- and disaccharides it appears that such a detachment of H from C_1 is highly probable in these compounds.

Supplementary EPR lines begin to appear more distinctly as the temperature of the irradiated samples rises to a level at which the OH and e_{st} radicals disappear [239]. The spectra of the irradiated samples generally differ in width as well as in appearance. The supplementary components appear both outside the doublet and in the center of the EPR spectrum. As the temperature rises the EPR spectra gradually become uniform; the radicals quickly vanish at 213–223°K, or in polysaccharide solutions at higher temperatures [145, 205].

The EPR spectra do not yet enable definitive identification of the radicals formed in irradiated frozen aqueous solutions of polysaccharides. The main difficulty in the interpretation of the EPR spectra of polysaccharides and their low molecular weight analogs lies in the great diversity of radicals generated in each of these carbohydrates, which adopt various structural forms in solution [239], and also in the instability of the radicals even at low temperatures.

We are not aware of any reliable determination of the structure of the primary radicals formed in the radiolysis of polysaccharides on the basis of the EPR spectra. Studies [240–242] of dry preparations of several polysaccharides (based on glucose) irradiated with γ-quanta or UV light at room temperature have revealed the following three types of radicals on the basis of the EPR spectra (the radicals formed under such conditions may be secondary): 1) an RO radical (singlet, $\Delta H_{max} = 19 \pm 2$ gauss), formed by cleavage of the O–H bond; 2) a radical with a doublet HFS ($\Delta H \approx 16$ gauss), resulting from the interaction of the unpaired electron with the proton of the hydrogen bond; 3) an RO_2 radical (a narrow asymmetrical singlet ($\Delta H_{max} = 8 \pm 1$ gauss), formed in the presence of oxygen as a result of cleavage of the C–O bond between the monomer units ($E_{bond} \geqslant 3.71$ eV) and the reaction described by Sharpatyi [23]. Further data, especially on low-temperature radiolysis, are necessary to understand the mechanism of the appearance and transformation of radicals in the above-mentioned polysaccharides.

In order to establish the nature of the radicals formed in frozen solutions of carbohydrates on the basis of their spectra, Nikitin et al. [236, 243] studied the radiolysis of frozen aqueous solutions of the crystallohydrate of meso-inositol and of dry preparations of this compound, a cyclic polyol which differs from other hexoses by the absence of oxygen in the ring. Assuming firstly, like Ueda [244], that the radicals generated by the irradiation of meso-inositol at 77°K retain essentially the structure of the original molecule, and accepting that the values of the splitting ΔH of the equatorial β-proton, the axial β-protons and the α-protons of meso-inositol are respectively 0, 36 and 18 gauss [245], the EPR spectrum obtained was interpreted as a superimposition of the lines originating from radicals formed by the detachment of H atoms from each of the carbon atoms – namely, a doublet (1:1, $\Delta H \approx 30$ gauss) and a triplet (1.2:1, $\Delta H \approx 30$ gauss) [236, 243]. The observed change in the EPR spectra as a result of the thermal annealing of radicals is explained [243] by the following scheme:

Norman et al. [245], on the basis of a study of the formation and conversion of radicals in liquid aqueous solutions of meso-inositol using the flow technique, conclude that such a dehydration of the radical can take place. Another possible type of transformation of the radical [243] is its isomerization, as a result of which the free valence migrates from C_α to the oxygen atom of the hydroxyl group (H migrates to C_α). Owing to this process, the line corresponding to the radical $R\dot{O}$ appears in the EPR spectrum at a g factor equal to 2 as the samples are gradually defrosted.

These conclusions on the formation and conversion of radicals in meso-inositol solutions are fully applicable to the radiolysis of solutions of mono-, di-, and polysaccharides. It may be assumed in general that the differences detected in the EPR spectra of carbohydrates of identical composition result from differences in the degree of overlapping of the orbits of the unpaired electron and the electrons of the C–H bond, even though the primary radicals formed from the dissolved carbohydrates may be identical. This means that the radical producing a doublet HFS, which is produced in greatest yield in polysaccharide solutions, results from the detachment of an H atom from C_1, while the radicals present at lower yields are formed by the detachment of an H atom from other carbon atoms.

The yield of each radical in any compound depends, of course, on the steric factor and the influence of heteroatoms on the mobility of the different hydrogen atoms in the monomer unit. The observed differences between the changes in the EPR spectra of irradiated carbohydrate samples of identical composition, whether in the course of defrosting or during photoannealing of the radicals, reflect also differences in the stability of the primary radicals formed from the dissolved carbohydrate, this stability being in turn a function of the structure of the radical; hence the different rates of disappearance of the radicals with a rise in temperature.

The structure of the primary radicals determines their stability, reactivity and other chemical properties, and in the final analysis, the mechanism of their transformation. Thus, the irradiation of polysaccharides [145], monosaccharides [147, 231, 246], disaccharides [146, 246], and glycosides [235] of identical composition but different structure, under uniform conditions (liquid or frozen solutions; in the latter case the products were analyzed after defrosting), leads to radiolytic products differing in yield and sometimes even in composition. For example, the composition of the radiolytic products obtained by the irradiation of frozen solutions of d-ribopyranose (a monomer with a stressed structure owing to the orientation of the hydroxyl groups in the molecule) depends on the temperature of radiolysis. Thus, irradiation of solutions at 195°K yields ribose epimers with a less stressed structure. This process does not occur if the radiolysis is carried out at room temperature.

These findings can be explained by assuming that the trend of the processes involved depends not only on the structure but also on the mobility of the radicals originating from the solute and those from water under the particular conditions of irradiation. Another example of the effect of the structure of the radical on its chemical properties can be found in studies of the photochemical properties of the radicals originating from mono- and disaccharides. Thus, experiments involving the UV illumination of irradiated frozen solutions of cellobiose (4-glycoside)-glucose and frozen solutions of glucose have shown that a formyl radical appears in the former but not in the latter [19, 158, 159].

According to the available data on the radiolysis of frozen aqueous solutions of polysaccharides and their low molecular weight analogs, the appearance and transformation of radicals in these systems can be outlined as follows.

Radicals from the solute are formed according to reactions (1.5) and (1.7):

$$RH + H(OH) = R + H_2(H_2O),$$

where the radicals R result mainly from the detachment of H from the C_1 and C'_1 atoms of the carbohydrate; there is a lesser probability that radicals will be formed by the detachment of an H atom from other carbons of the sugar ring. The radiolytic products of water which do not participate in the reaction at 77°, namely, e_{aq} and some of the OH radicals, are stabilized in the matrix as e_{st} and OH. The following processes take place on defrosting of the sample:

$$OH + RH = H_2O + R;$$

$$OH + e_{st} = OH^-;$$

$$e + R = R^- \xrightarrow{+H^+} RH.$$

With a rise in temperature the radicals originating from the solute may isomerize within the same ring into radicals of the RO·type, or by a transfer of the free valence to the neighboring ring, accompanied by cleavage of the glycoside bonds, they can also undergo dehydration or participate in disproportionation reactions.

The mechanism of transformation of the radicals formed in aqueous solutions of polysaccharides in the presence of molecular oxygen or other additives such as radioprotectors containing sulfhydryl groups includes also reactions of the type $R + O_2 = RO_2$ with subsequent formation of hydroperoxides and $R + H-SR' = RH + \cdot SR'$. As in the case of protein-type biopolymers, it is assumed [227] that the radioprotector molecule manifests its protective properties by forming a complex with the biopolymer molecule, thus allowing migration of the free valence from the macroradical to the radioprotector. In the case of Sephadex and penicillamine solutions, a protective effect is obtained in the presence of one molecule of penicillamine per 13 glycoside monomer units [205].

The available data on the radiolysis of frozen aqueous solutions of biopolymers thus indicate that the processes detected by radiochemical methods undoubtedly perform an important if not crucial function in radiation-induced lesions of biopolymers in real, native systems.

CONCLUSION

To sum up, we shall try to formulate the main achievements of radiochemistry with respect to the systems discussed and outline the trends of future research. The data obtained by the radiochemical analysis of liquid and frozen aqueous solutions are in mutual agreement and point to an indirect effect of radiation in these systems. This is evident from the regularities detected in the radiochemistry of aqueous solutions and the participation of the radiolytic products of water; the involvement of OH radicals and hydrated electrons in the reactions and also the sites at which these products attack the biopolymer molecule have been demonstrated for a number of objects. Yet a complete picture of the effect of radiation on such a complex system as the cell cannot be obtained without an analysis of the processes combining the direct and indirect effects of radiations — first of all the migration of the excitation energy along the solvent molecules into the hydrate sheaths of the functional groups of the biopolymers, then the breakdown of the water molecules forming these sheaths into radicals which react with the substrate (according to Proskurnin).

Hence the importance of research into the structural properties of the cell components, the structure of this system as a whole, and the effect of structural factors on the radiolytic transformations of the substrate. This approach is essential for the practical solution of such problems as the protection and sensitization of the breakdown. It is evident from the available data on the chemical components of the cell that the magnitude of the radiolytic protection or radiosensitization with respect to irradiation of the system depends on the ability of the additive to affect the canals transmitting the excitation energy in the irradiated system, i. e., the sections of the cell consisting of dilute or concentrated aqueous solutions. The mode of action of radioprotectors and radiosensitizers may differ in these two cases.

It is evident from the available data on the radiolysis of aqueous solutions of cell biopolymers that radioprotectors, like radiosensitizers, can function by means of several mechanisms including competition with the substrate for the radiolytic products of water; capture of charges and excitation energy by the radioprotector introduced (during the formation of damaged states of the substrate); capture of the electron charge by the radioprotector molecule and inhibition of free-radical reactions (as described by Emanuel) during the subsequent chemical transformation of the radiolytic products of the substrate. Radiosensitizers evidently function in an analogous way, although in contrast to radioprotectors they yield radiolytic products which are active toward the substrate.

Examination of the available data on the radiochemistry of compounds of biological origin and the results of radiobiological research reveals a great gap between, on the one hand, the existing knowledge of the mechanisms of radiation at the level of the animal organism and those at the primary

stages of radiolysis in model systems at the molecular level and, on the
other hand, the mechanisms controlling the development of radiation damage
in the intermediate region linking the primary stages of radiolysis with the
final biochemical and physiological manifestations of the effect of irradiation
on the live organism. We believe, therefore, that research in the near
future will be focused on the radiochemical effects of radiation in this intermediate region, using objects of gradually increasing complexity: aqueous
solutions of biopolymers → solutions containing several cell components →
natural complexes → organelles → the cell. By this means, it should be
possible to define the mechanism governing the development of radiation
damage in physicochemical terms and correlate it with the biological and
physiological phenomena forming the subject of radiation pathology.

BIBLIOGRAPHY

1. K u z i n , A.M. Radiation Biochemistry. — Moscow, Izd. AN SSSR, 1962. (Russian)
2. B a c q , Z. and P. A l e x a n d e r .Fundamentals of Radiation Biology. Edited by Ya.M.Varshavskii,
        E.Ya.Graevskii, and M.N.Meisel.— Moscow, I.L., 1963. (Russian translation)
3. R o m a n t s e v , E.F. Radiation and Chemical Protection. 2nd edition.— Moscow, Atomizdat, 1968. (Russian)
4. A m i r a g o v a , M.I. et al. Primary Radiobiological Processes.— Moscow, Atomizdat, 1964. (Russian)
5. P i s a r e v s k i i , A.N. et al. Introduction to Radiation Biophysics.— Minsk, "Vysshaya Shkola," 1968. (Russian)
6. B a c q , Z. Chemical Protection against Ionizing Radiation. Edited by A.M.Kuzin.— Moscow, Atomizdat,
        1968. (Russian translation)
7. S p i t k o v s k i i , D.M. et al. Radiation Biophysics of Nucleoprotein.— Moscow, Atomizdat, 1969. (Russian)
8. S h a r p a t y i , V.A.— Uspekhi Khimii 30 (1961), 645.
9. S h a r p a t y i , V.A.— Thesis. Moscow, 1960.
10. K r a l i č , I. and C.N. T r u m b o r e . — J. Am. Chem. Soc. 87 (1965), 2547.
11. K r a l i č , I. The Chemistry of Ionization and Excitation.— London, Taylor and Francis (1967), 303.
12. W a r d , J. and L.S.M y e r s . — Radiat. Res. 26 (1965), 483.
13. A n b a r , M. et al.— J. Chem. Soc. 13 (1966), 742.
14. B a l a z s , E.A. et al.— Radiat. Res. 31 (1967), 243.
15. G r e e n s t o c k , C.L. and J.W.H a n t.— Radiat. Chem., 1, adv. chem. ser. 81 (1968), 397
16. S c h o l e s , G.S. et al. Pulse Radiolysis.— London—New York, Academic Press (1965), 151.
17. K a l k w a r f , D.R. — In: Radiation Effects in Physics, Chemistry and Biology. Edited by D.E. Grodzenskii,
        P.D.Gorizontov. Moscow, Atomizdat (1965). 295. (Russian translation)
18. K a l k w a r f , D.R.— Nucl. Sci. Abst. 17, No.13966. 1963.
19. K o c h e t k o v , N.K. et al.— Khimiya Vysokikh Energii 2 (1968), 566.
20. Z a k a t o v a , N.V. and V.A.S h a r p a t y i.— Izv. AN SSSR, chem. ser. 7 (1968), 1642.
21. Z a k a t o v a , N.V. et al.— Izv. AN SSSR, chem. ser. 7 (1969), 1633.
22. Z a k a t o v a , N.V. et al.— Khimiya Vysokikh Energii 3 (1969), 365.
23. S h a r p a t y i , V.A. et al.— Khimiya Vysokikh Energii 2 (1968), 473.
24. V o l k e r t , W.A. and R.R.K u n t z.— J. Phys. Chem. 72 (1968), 3394.
25. A d a m s , G.E. et al. Pulse Radiolysis.— L.—N.Y., Academic Press (1965), 131.
26. D a v i e s , J.V. et al.— Ibid., p.181.
27. P h i l i p s , G.O. et al.— J. Chem. Soc. B (1966), 194.
28. S h a r p a t y i , V.A. and N.V.Z a k a t o v a.— Inform. Byull. "Radiobiologiya," No.12 (1969), 63.
29. B a x e n d a l e , J.H. and A.K.K h a n.— Int. J. Radiat. Phys. Chem. 1 (1969), 11.
30. S h a r p a t y i , V.A. and Yu.N.M o l i n.— Proc. 2nd All-Union Conf. on Radiation Chemistry, p.141.
        Moscow, AN SSSR, 1962. (Russian)
31. S h a r p a t y i , V.A.— Proc. Tihany Symp. on Radiation Chemistry, p.463. Budapest, Publ. House Hungar.
        Acad. Sci., 1964.
32. S h a r p a t y i , V.A. et al.— In: Fizicheskie Problemy Spektroskopii 2, p.100. Moscow, AN SSSR, 1963.
33. F e l ' , N.S. et al.— Khimiya Vysokikh Energii 2 (1968), 376.

34. Sharpatyi, V.A. and N.V.Zakatova.— Izv. AN SSSR, chem. ser. (1968), 2197.
35. Neta, P. and L.M.Dorfman.— Radiat. Chem., adv. chem. ser. 81 (1968), 222.
36. Rabani, J. and M.S.Matheson.— J. Phys. Chem. 70 (1966), 761.
37. Myers, L.S. et al.— Radiat. Chem. 1, adv. chem. ser. 81 (1968), 345.
38. Scholes, G. et al.— Chem. Comm. 1D (1969), 17.
39. Davies, J.V. et al.— Int. J. Radiat. Biol. 14 (1968), 19.
40. Chrysochoos, J.— Radiat. Res. 33 (1968), 465.
41. Rabani, J.— Radiat. Chem. 1, adv. chem. ser. 81 (1968), 131.
42. Gordon, S. et al.— Discuss. Faraday Soc. 36 (1963), 193.
43. Gordon, S. et al.— J. Phys. Chem. 68 (1964), 1262.
44. Armstrong, W.A.— Can. J. Chem. 47 (1969), 3737.
45. Brodskaya, G.A. and V.A.Sharpatyi.— Khimiya Vysokikh Energii 2 (1968), 184.
46. Sworski, T.J.— Radiat. Res. 2 (1965), 26.
47. Byakov, V.M. and B.V.Ershler.— Doklady AN SSSR 154 (1964), 669.
48. Mahlman, H.A.— J. Chem. Phys. 32 (1960), 601.
49. Appleby, A. The Chemistry of Ionization and Excitation, p.269.— London, Taylor and Francis, 1967.
50. Mahlman, H.A. and T.J.Sworski.— Ibid., p.259.
51. Brodskaya, G.A.— Thesis. Tashkent, 1968.
52. Brodskaya, G.A. and V.A.Sharpatyi.— Khimiya Vysokikh Energii 2 (1968), 254.
53. Brodskaya, G.A. and V.A.Sharpatyi.— Zh. Fiz. Khim. 48 (1969), 2390.
54. Sharpatyi, V.A. et al.— Zh. Fiz. Khim. 39 (1965), 232.
55. Scholes, G. and M. Simic.— Biochim. biophys. Acta 166 (1968), 255.
56. Scholes, G. et al.— Nature 178 (1956), 157.
57. Scholes, G. Progress in Biophys. and Mol. Biol.— Pergamon Press (1963), 13, 61.
58. Nofre, C. and A.Cier.— Bull. Soc. chim. Fr. 4 (1966), 1326.
59. Pikaev, A.K. The Solvated Electron in Radiation Chemistry.— Moscow, "Nauka," 1969. (Russian)
60. Balazs, E.A. et al.— Radiat. Res. 31 (1967), 243.
61. Braams, R. Pulse Radiolysis.— L.–N.Y., Academic Press (1965), 171.
62. Braams, R.— Radiat. Res. 27 (1966), 319.
63. Adams, G.E. et al.— Trans. Faraday Soc. 65 (1969), 732.
64. Adams, G.E. Book of Abstracts.— Roma, Istituto Superiore di Sanita (1969), 7.
65. Greenstock, C.L. et al.— Ibid., p.55.
66. Emmerson, P.T. and R.L.Willson.— Ibid., p.42.
67. Emmerson, P.T. and R.L.Willson.— J. Phys. Chem. 72 (1968), 3669.
68. Proskurnin, M.A. et al.— Uspekhi Khimii 24 (1955), 584.
69. Proskurnin, M.A.— In: "Problemy Fizicheskoi Khimii," No.1, p.48. Moscow, Goskhimizdat, 1958.
70. Orekhov, V.D. et al.— Proc. 1st All-Union Conf. on Radiation Chemistry, p.100. Moscow, AN SSSR, 1957. (Russian)
71. Proskurnin, M.A.— Doklady AN SSSR 135 (1960), 1446.
72. Proskurnin, M.A. and V.A.Sharpatyi.— Zh. Fiz. Khim. 34 (1960), 2126.
73. Sharpatyi, V.A. et al.— Doklady AN SSSR 122 (1958), 852.
74. Krushinskaya, N.P.— Thesis. Moscow, 1970.
75. Granovskaya, M.A. et al.— Radiobiologiya 5 (1965), 633.
76. Daile, W.M. and J.V.Davies.— Biochem. J. 48 (1951), 129.
77. Brudička, R. and Z.Spurný,— Collection 4 (1958), 561.
78. Whitcher, S.L. et al.— Nucleonics 11 (1953), 30.
79. Fojtik, A. et al.— Collection 11 (1969), 3623.
80. Rowbottom, J.— J. Biol. Chem. 221 (1955), 877.
81. Swallow, A.J.— J. Chem. Soc. (1952), 1334.
82. Lal, M., D.A.Armstrong, and M.Wieser.— Radiat. Res. 37 (1969), 246.
83. Yarovaya, S.M.— Thesis. Moscow, 1967.
84. Kochetkov, N.K. et al.— Zhurnal Obshchei Khimii 35 (1965), 268.
85. Kudryashov, L.I. et al.— Zhurnal Obshchei Khimii 41 (1971), 298.
86. Swallow, A.— In: Radiation Chemistry of Organic Compounds. Edited by V.L.Karpov. Moscow, I.L. (1963), p.218. (Russian translation)
87. Khenokh, M.A. and E.M.Lapinskaya.— Doklady AN SSSR 110 (1956), 125.

88. D a l e , W.M. — In: Ionizing Radiation and Cellular Metabolism. Moscow, I.L. (1958), 43. Edited by
     M.N.Meisel'. (Russian translation)
89. A r m s t r o n g , K.C. and A.C h a r l e s b y.— Int. J. Radiat. Biol. 12 (1967), 523.
90. G o r i n , G. et al.— Int. J. Radiat. Biol. 15 (1969), 33.
91. G o r i n , G. et al.— Ibid., p.23.
92. G o r i n , G. et al.— Int. J. Radiat. Biol. 16 (1969), 93.
93. D o s e , K. et al.— Biophysik 3 (1966), 202.
94. A d a m s , G.E. et al.— Int. J. Radiat. Biol. 16 (1969), 333.
95. A l d r i c h , J.E. and R.B.C u n d a l l.— Ibid., p.343.
96. L y n n , V.R. and O r p e n Gail.— Radiat. Res. 39 (1969), 15.
97. D a l e , W.M. et al.— Phil. Trans. R. Soc. A 242 (1949), 33.
98. A f i f i , F. and M.J.H a i s s i n s k y.— Pucheault, Compt. rend.250 (1960), 118.
99. H a y m o n d , H.R. and S.M.A.H a s a n.— Radiat. Res. 39 (1969), 482.
100. R a b a n i , J. and G.S t e i n.— Radiat. Res. 17 (1962), 327.
101. S t e i n , G. and M.T o m k i e w i c z.— Radiat. Res. 43 (1970), 25.
102. L a b o u t , J.J.M.— Nucl. Sci. Abstr. 24, No.20790 (1970), 2054.
103. S u t t o n , H.C.— Biochem. J. 64 (1956), 447.
104. P r i s t u p a , A.I. et al.— Izv. AN SSSR, chem. ser., No.2 (1970), 488.
105. S a p e z h i n s k i i , I.I. See p.95.
106. S i l a e v , Yu.V.— Proc. Sympos. of the Moscow Society of Naturalists. In: "Sverkhslabye svecheniya v
     biologii," p.56. Moscow, MGU, 1969. (Russian)
107. P h i l l i p s , G.O. and N.W.W o r t h i n g t o n. — Radiat. Res. 43 (1970), 34.
108. C h a m b e r s , K.W. et al.— Trans. Faraday Soc. 66 (1970), 142.
109. B l o k , J. and W.S.D.V e r h e y.— Radiat. Res. 34 (1968), 689.
110. S c h o l e s , G. and J.W e i s s.— Nature 173 (1954), 267.
111. B a r r o n , E.S.G. and V.F l o o d.— J. Gen. Physiol. 33 (1950), 229.
112. S h a p i r a , R. et al. Radiat. Res. 7 (1957), 22.
113. A n d e r s o n , D.R. and B.I.J o s e p h.— Radiat. Res. 10 (1959), 507.
114. K a m a l , A. and M.G a r r i s o n.— Nature 206 (1965), 1315.
115. Z h i z h i n a , G.P.— Thesis. Moscow, 1968.
116. Z a k a t o v a , N.V. and V.A.S h a r p a t y i.— Synopses of Reports All-Union Conf. of Young Scientists on
     Radiation Chemistry and Radiation Biology 24—28 November, 1969, Obninsk, p.107. (Russian)
117. Z a k a t o v a , N.V. and V.A.S h a r p a t y i.— Reports All-Union Conf. of Young Scientists on Radiobiology
     15—17 September, 1969, Pushchino-na-Oke, p.26. (Russian)
118. E v a n s , D.F.— J. Chem. Soc. D, No.7 (1969), 367.
119. E m a n u e l' , N.M.— Trudy Moscow Soc. Naturalists (MOIP) 7 (1963), 73.
120. B u r l a k o v a , E.B. et al.— Reports Conf.25—28 February, 1957 on the Problem of "Biokhimicheskie i
     fiziko-khimicheskii osnovy biologicheskogo deistviya radiatsii," p.16, Moscow, MGU, 1957.
121. E m a n u e l' , N.M. et al.— Doklady AN SSSR 131 (1960), 1451.
122. E m a n u e l' , N.M. et al. Physikalische Chemie biogener Makromoleküle.— 2 Jenaer Symposium, 1963,
     p.349. Berlin, Akad. Verlag, 1964.
123. E m a n u e l' , N.M. et al.— Izv. AN SSSR, biol. ser., No.2 (1966), 183.
124. K r u g l y a k o v a , K.E.— Thesis. Moscow, 1969.
125. E r s h o v , B.G. and A.K.P i k a e v.— Zh. Fiz. Khim. 41 (1967), 2573.
126. F l o u r n o y , J.M. et al.— J. Chem. Phys. 36 (1962), 2229.
127. S i e g e l , S. et al.— J. Chem. Phys. 34 (1961), 1782.
128. S h a r p a t y i , V.A.— Uspekhi Khimii 32 (1963), 737.
129. M o o r t h y , P.N. and J.J.W e i s s.— J. Chem. Phys. 42 (1965), 3121.
130. M o o r t h y , P.N. and J.J.W e i s s.— Ibid., p.3127.
131. E r s h o v , B.G. et al.— Int. J. Radiat. Phys. Chem. 3 (1971), 7.
132. J o n e s , W.M.— J. Chem. Phys. 20 (1952), 1974.
133. G h o r m l e y , J.A.— J. Chem. Phys. 24 (1956), 1111.
134. G r o s s w e i m e r , L.I. and M.S.M a t h e s o n. —J. Chem.Phys.22 (1954), 1514.
135. S h a r p a t y i , V.A. et al.— Khimiya Vysokikh Energii 3 (1969), 469.
136. B e t t i n a l i , C. and G.F e r r a r e s s o.— J. Chem. Phys. 48 (1968), 517.
137. S h a r p a t y i , V.A. et al.— Doklady AN SSSR 180 (1968), 412.

138. Zimbrick,J. and L.Kevan.— J. Am. Chem. Soc. 88 (1966),3678.
139. Zimbrick,J. and L.Kevan.— J. Chem. Phys. 47 (1967),2364.
140. Zimbrick,J. and L.Kevan.— Ibid.,p.5000.
141. Sharpatyi,V.A. et al.— Izv. AN SSSR, chem. ser., No.3 (1970), 702.
142. Sharpatyi,V.A. et al.— Radiobiologiya 8 (1968),3.
143. Sharpaty, V.A.— Studia biophysica 8 (1968),187.
144. Sharpatyi,V.A.— 7th Intern. Sympos. on Chemistry of Natural Compounds,p.226. Riga, "Zinatne," 1970. (Russian)
145. Sharpatyi,V.A.— Doklady AN SSSR 181 (1968),655.
146. Emanuel',N.M. et al.— Doklady AN SSSR 177 (1967),1142.
147. Nadzhimiddinova,M.T. and V.A.Sharpatyi.— Doklady AN SSSR 180 (1968),909.
148. Sharpatyi,V.A. and M.N.Sultankhodzhaeva.— Izv. AN SSSR, chem. ser., No.5 (1969),1183.
149. Bel'kovich,T.M.— Ibid.,p.2652.
150. Sharpatyi,V.A. and Yu.N.Molin.— Zh. Fiz. Khim. 35 (1961),1465.
151. Sharpatyi,V.A. and K.G.Yanova.— Zh. Fiz. Khim. 37 (1963),948.
152. Siegel,S. et al.— J. Chem. Phys. 30 (1959),1623.
153. Köhnlein,W. and D.Schulte-Frohlinde.— Radiat. Res. 38 (1969),173.
154. Sharpatyi,V.A.— In: "Deistvie radioaktivnykh izluchenii na veshchestvo," p.54. Tashkent,FAN,1970.
155. GIICHI Yoshii IV Congrès Intern. de Radiobiologie et de phys.-chim. de Rayon, Evian,1970,Livre des Résumés, No.928,p.236.
156. Sanaev,B. et al.— Zh. Fiz. Khim. 39 (1965),2510.
157. Elliot,W.R.— Science 157 (1967),558.
158. Nadzhimiddinova,M.T. et al.— Doklady AN SSSR 182 (1968),1354.
159. Nadzhimiddinova,M.T. and V.A.Sharpatyi.— Proc. Sympos. on "The Effect of Ionizing Radiation on Proteins and Nucleic Acids. Molecular Mechanisms of Protection. 4—8 July,1968," p.24.— Kiev, "Naukova Dumka," 1968. (Russian)
160. Sharpatyi,V.A. et al.— Ibid.,p.301.
161. Sharpatyi,V.A.— Synopses of Reports of the 2nd Republ. Conf. Mechanism of Biological Effect of Ionizing Radiation. 3—6 July,1969,p.335. Lvov,1969. (Russian)
162. Müller,A. and W.Köhnein.— Int. J. Radiat. Biol. 8 (1964),121,131,141.
163. Müller,A.— Prog. Biophys. Molec. Biol.,17 (1967),101.
164. Golikov,V.P. et al.— Khimiya Vysokikh Energii 2 (1968),449.
165. Golikov,V.P. et al.— Radiobiologiya 9 (1969),346.
166. Lenherr,A.D. and M.G.Ormerod.— Biochim. biophys. Acta 166 (1968),298.
167. Miura,M. et al.— Bull. Chem. Soc. Japan 39 (1966),1432.
168. Hüttermann,J. and A.Müller.— Z. Naturf. 246 (1969),463.
169. Hüttermann,J. and A.Müller.— Radiat. Res. 38 (1969),248.
170. Norther,J. et al.— J. Chem. Phys. 37 (1962),2488.
171. Kevan,L. et al.— J. Am. Chem. Soc. 86 (1964),771.
172. Srinivasan,V.T. et al.— Int. J. Radiat. Biol. 15 (1969),89; 17 (1970),577.
173. Lenherr,A.D. and M.G.Ormerod.— Nature 225 (1970),1302.
174. Nazhat,N.B. and J.J.Weiss.— Trans. Faraday Soc. 66 (1970),1302.
175. Sevilla,M.D.— J. Phys. Chem. 74 (1970),805.
176. Herak,J.N. and W.Gordy.— Proc. Natn. Acad. Sci. U.S.A. 54 (1965),1287.
177. Herak,J.N. and W.Gordy.— Proc. Natn. Acad. Sci. U.S.A. 55 (1966),1373.
178. Heller,H.C. and T.Cole.— Proc. Natn. Acad. Sci. U.S.A. 54 (1965),1486.
179. Holmes,D.E. et al.— Int. J. Radiat. Biol. 12 (1967),225,415.
180. Holroyd,R.A. and J.W.Glass.— Radiat. Res. 35 (1968),510.
181. Holroyd,R.A. and J.W.Glass.— Radiat. Res. 39 (1969),758.
182. Jensen,H. and T.Henriksen.— Acta chem. scand. 22 (1968),2263.
183. Henriksen,T.—J. Chem. Phys. 50 (1969),4653.
184. Henriksen,T.— Radiat. Res. 40 (1969),11.
185. Benson,B. and W.Snipes.— Int. J. Radiat. Biol. 15 (1969),583.
186. Abagyan,G.V. and P.Yu.Butyakin.— Biofizika 14 (1969),785.
187. Sanner,T.— Radiat. Res. 25 (1965),586.
188. Sharpatyi,V.A. et al.— Doklady AN SSSR 203 (1972),147.

189. Nice,E.G. and D.Rorke.– Int. J. Radiat. Biol. 15 (1969),207.

190. Hüttermann,J. and A.Müller.– Int. J. Radiat. Biol. 15 (1969),297.

191. Nice,E.G. and D.Rorke.– Int. J. Radiat. Biol. 15 (1969),197.

192. Biemann,K. and J.M.McClosky.– J. Am. Chem. Soc. 84 (1962),2005.

193. Milinchuk,V.K. et al.– Radiobiologiya 9, No.5 (1969),659.

194. Nicolau,C. et al.– Biochim. biophys. Acta 174 (1969),413.

195. Ginzburg,S.F. and M.S.Postnikova.– Khimiya Vysokikh Energii 3 (1969),381.

196. Postnikova,M.S. et al.– Biofizika 14 (1969),927.

197. Wheaton,R.E. and M.Ormerod.– Trans. Faraday Soc. 65 (1969),1638.

198. Copeland,E.S. and H.M.Swartz.– Int. J. Radiat. Biol. 16 (1969),293.

199. Krivenko,V.G. et al.– Khimiya Vysokikh Energii 4 (1970),49.

200. Henriksen,T.– J. Chem. Phys. 38 (1963),1962.

201. Sharpatyi,V.A. et al.– Trudy Mosk. Obshch. Ispyt. Prir. 16 (1966),110.

202. Henriksen,T.– Radiat. Res. 17 (1962),158.

203. Pihl,A. et al.– Acta chem. scand. 17 (1963),2124.

204. Lassmann,G.– Z. phys. chem. 225 (1964),409.

205. Sanner,T. et al.– Radiat. Res. 32 (1967),463.

206. Liming,F.G. and W.Gordy.– Proc. Natn. Acad. Sci. U.S.A. 60 (1968),740,794.

207. Liming,F.G.– Radiat. Res. 39 (1969),252.

208. Sharpatyi,V.A. et al.– Doklady AN SSSR 155 (1964),626.

209. Sharpatyi,V.A. and A.N.Pravednikov.– J. Polym. Sci. CH, No.16 (1967),1599.

210. Sharpatyi,V.A. et al.– Doklady AN SSSR 157 (1964),660.

211. Taniguchi,H. et al.– J. Phys. Chem. 72 (1968),1926.

212. Sultanov,A. and V.A.Sharpatyi.– Khimiya Vysokikh Energii 1 (1967),541.

213. Brodskaya,G.A. and V.A.Sharpatyi.– Khimiya Vysokikh Energii 1 (1967),336.

214. Usatyi,A.F. et al.– Khimiya Vysokikh Energii 2 (1968),444.

215. Meybeck,A. and J.Windle.– Radiat. Res. 40 (1969),263.

216. Panin,V.I. et al.– Khimiya Vysokikh Energii 3 (1969),248.

217. Copeland,E.S.– Int. J. Radiat. Biol. 16 (1969),113.

218. Henriksen,T. et al.– Radiat. Res. 18 (1963),147.

219. Pulatova,M.K.– Thesis. Moscow,1964.

220. Hong,S.J. et al.– Radiat. Res. 35 (1968),509.

221. Henriksen,T.– Scand. J. Clin. Lab. Invest.,22,Suppl. 106 (1968),7.

222. Pasoyan,V.G. et al.– Biofizika 13 (1968),716.

223. Pasoyan,V.G. et al.– Biofizika 15 (1970),12.

224. Kayushin,L.P. et al.– Study of the Paramagnetic Centers of Irradiated Proteins.– Moscow, "Nauka," (1970),174. (Russian)

225. Krivenko,V.G. et al.– Khimiya Vysokikh Energii 4 (1970),49.

226. Willix,R.L.S. and W.M.Garrison.– Radiat. Res. 32 (1967),452.

227. Pihl,A. et al.– Radiat. Res. 35 (1968),235.

228. Copeland,F.S. et al.– Radiat. Res. 35 (1968),437.

229. Sharpatyi,V.A. and K.A.Korotchenko.– Synopses of Reports of the All-Union Sympos. on the Phys. and Chem. Properties of Starch and its Products,18–20 June,1968. Moscow,Vses. Tsentr. Sovet Profess. Soyuzov.– Vses. Sovet Nauch.– Tekh. Obshchestv, (1968),7. (Russian)

230. Kochetkov,N.K. et al.– Zh. Obshch. Khim. 35 (1965),2246.

231. Nadzhimiddinova,M.T. and V.A.Sharpatyi.– Proc. Conf. Young Scientists on Radiation Chemistry. Minsk,1967. (Russian)

232. Nadzhimiddinova,M.T. et al.– Synopses of Reports of All-Union Sympos. on the Phys. and Chem. Properties of Starch and its Products,18–20 June 1968. Moscow,Vses. Tsentr. Sovet Profess. Soyuzov.– Vses. Sovet Nauch.–Tekh. Obshchestv (1968), 7. (Russian)

233. Kochetkov,N.K. et al.– Doklady AN SSSR 183 (1968),376.

234. Sharpatyi,V.A.– Khimiya Vysokikh Energii 2 (1968),286.

235. Kudryashov,L.I. et al.– Zh. Obshch. Khim. XL (1970),1133.

236. Nikitin,I.V. et al.– Synopses of Reports of the 2nd Republ. Conf. on Mechanisms of the Biological Effect of Ionizing Radiation,3–6 July. Lvov (1969),216.

237. Yarovaya,S.M. et al.– Ibid.,p.346.

238. Nadzhimiddinova,M.T.— Thesis. Tashkent,1969.

239. Kochetkov,N.K. Carbohydrate Chemistry.— Moscow, "Khimiya," 1967.

240. Gol'din,S.I. and S.V.Markevich.— Vest. Akad. Nauk BSSR,chem. ser., No.6 (1970),47.

241. Gol'din,S.I. et al.— Vest. Akad. Nauk BSSR,chem. ser., No.2 (1971),57.

242. Gol'din,S.I. and S.V.Markevich.— Khimiya Vysokikh Energii 5 (1971),463.

243. Nikitin,I.V. et al.— Doklady AN SSSR 190 (1970),635.

244. Ueda,H.— J. Phys. Chem. 67 (1963),2185.

245. Norman,R.O.C. and R.J.Pritchett.— J. Chem. Soc. B (1967),1329.

246. Baugh,P.J. et al.— Nature 221 (1969),1138.

Chapter 2

THE RADIOLYSIS OF DILUTE AQUEOUS SOLUTIONS OF MONOSACCHARIDES AND THEIR DERIVATIVES

INTRODUCTION

Recent chemical and biochemical research has shed new light on the role of carbohydrates in vital processes. Without entering into a detailed discussion of the various ways in which carbohydrates participate in biochemical processes, we can state that these compounds, like nucleic acids, proteins and lipids, form high molecular weight complexes incorporated as constituents in submolecular structures which are highly susceptible to physicochemical factors, including radiation.

Little is known as yet about the mechanism of the transformation of carbohydrates under the influence of radiation, in particular, the sequence of conversions, the structure of the final products, etc. Yet these topics are crucial to an understanding of the mode of action of radiation on carbohydrate-containing biopolymers [1–3].

This chapter deals with the effect of ionizing radiations on dilute aqueous solutions of monosaccharides and their derivatives: a number of glycosides with different types of glycoside bonds; disaccharides, which can be regarded as the simplest building stones of natural polymers; and finally, compounds of the glucosamine type, one of the main components of compound biopolymers.

The radiolytic transformations of several carbohydrates were studied under comparable conditions in order to determine the relationship between their structure and behavior during the course of radiolysis. Most of the compounds tested were quite dilute $(5 \cdot 10^{-3} - 5 \cdot 10^{-2} \text{ M})$, so that a direct effect of the ionizing radiations could practically be ruled out and the chemical changes observed in the solutions could be attributed to reactions between the solute and radicals originating from the radiolysis of water (the indirect effect of radiation). Co^{60} at a dose rate of 800–300 rad/sec was used as the source of radiation.

THE STABILITY OF CARBOHYDRATES IN AQUEOUS SOLUTION TO IONIZING RADIATIONS

It was found that in most cases the concentration of the compound tested decreases linearly with the dose during the initial stage of radiolysis; an exponential relationship is observed if a wider range of doses is used (in disaccharide solutions continuing until the breakdown of about 60% of the compound).   The first-order nature of the reaction indicates that transformations of the carbohydrates under such conditions involve only radicals produced by the radiolysis of water − in this case, H, OH, and e_{aq}, the stationary concentration of which changes only slightly during the initial stage of the process.   The breakdown yield for the majority of the carbohydrates examined ranges from 2.5 to 3.5 molecules per 100 eV of absorbed energy, i.e., it is within the limits common to many organic compounds irradiated under similar conditions (Table 2.1).

TABLE 2.1. Breakdown yield during the radiolysis of 10^{-2} M solutions of carbohydrates irradiated in a nitrogen atmosphere, pH=7, dose range 0.2−0.8 Mrad

Compound	Yield, molecules/100 eV		References
	break-down	formation of monosaccharides (glucose)	
Glucose	2.4	−	[4]
Galactose	2.5	−	[4]
Mannose	2.8	−	[4]
α-Methyl-D-glucoside	2.6	0.1	[5]
β-Methyl-D-glucoside	3.2	0.1	[5]
β-Methyl-L-arabinoside	2.5	0.1 (arabinose)	[6]
α-Methyl-D-galactoside	5.0	0.1 (galactose)	[6]
β-Phenyl-D-glucoside	3.1	0.3	[5]
β-Benzyl-D-glucoside	3.4	1.0	[5]
β-p-Nitrophenyl-D-glucoside	2.7	1.0	[5]
Cellobiose	3.1	0.8	[7]
Maltose	3.3	1.0	[7]
Gentiobiose	4.0	1.2	[7]
Lactose	2.6	0.8 (glucose)	[7]
		0.4 (galactose)	[7]
D-Glucosamine base	7.3	−	[8]
N-Acetyl-D-glucosamine	3.3	−	[8]
β-Methyl-D-glucosaminide	3.3	−	[8]
β-Methyl-N-acetyl-D-glucosaminide	4.0	−	[8]
α-Methyl-N-acetyl-D-glucosaminide	2.5	−	[8]

It is noteworthy that in most cases the breakdown yield exceeds 2.7 molecules per 100 eV, i.e., it corresponds to more than half the yield of radicals originating from ionized water molecules under these conditions (5.4 molecules per 100 eV).   This means, on the one hand, that only one radical participates in the transformation of the original sugar molecule, i.e., the process does not include reactions such as

$$RH + OH \rightarrow R \cdot H_2O; \qquad (2.1)$$
$$R \cdot OH \rightarrow ROH. \qquad (2.2)$$

On the other hand, the fact that the breakdown yield exceeds $G(OH) = 2.7$ suggests that other radicals (H, e_{aq}) may take part in the transformations of the carbohydrates. In some cases (α-methylgalactoside, gentiobiose, glucosamine base), the breakdown yield is roughly equal to or slightly greater than the total yield of radicals formed. A study of the breakdown yield as a function of the concentration of the initial compounds was made in order to establish the causes of this phenomenon (Table 2.2).

TABLE 2.2. Effect of concentration on the breakdown yield (molecules/100 eV)

Compound	Concentration, M					References
	$5 \cdot 10^{-3}$	10^{-2}	$5 \cdot 10^{-2}$	10^{-1}	$6.7 \cdot 10^{-1}$	
Glucose	2.5	2.4	3.1	3.3	–	[4]
Mannose	2.3	2.8	3.1	3.7	–	[4]
Benzylglucoside	–	3.2	4.4	–	–	[9]
Cellobiose	–	3.1	–	–	–	[10]
Lactose	–	2.3	–	–	3.4	[10]
Glucosamine	–	7.3	440	21	–	[8]
N-Acetylglucosamine	–	5.5	31	9	–	[8]
β-Methylglucosaminide	–	3.3	6.0	–	–	[8]

It was found that the breakdown yield of mono- and disaccharides gradually increases with the concentration of these compounds to a maximum of 3.3–3.7. This may be attributed to a more complete involvement of the reductive component of radiolysis (H, e_{aq}) in the reaction and the capture of some of the OH radicals (which are most reactive toward the carbons of carbohydrates) from excited water molecules. Probably, both processes can take place. The participation of e_{aq} in the radiolysis of carbohydrates is evident from the marked differences between the breakdown yields of several isomeric glycosides and disaccharides. The increased breakdown yield in the radiolysis of solutions of methylgalactoside and gentiobiose probably results from the involvement of e_{aq} in the reaction since the rate constants of the reaction between the OH radical and carbohydrates are rather high (about $9-10 \cdot 10^9$ $M^{-1} sec^{-1}$) and largely independent of stereochemistry [11]. The increase in the breakdown yield of β-benzylglucoside is much greater and occurs within a narrow range of concentrations. Such a drastic increase in the breakdown yield is typical of processes which involve excited water molecules.

Radically different results are obtained in the radiolysis of solutions of glucosamine and N-acetylglucosamine. Here, the breakdown yield rises sharply with concentration up to $5 \cdot 10^{-2}$ M, i. e., up to 440 and 31 molecules for glucosamine and N-acetylglucosamine, respectively. Any further increase in the concentration causes a sharp decline in the yield. Such high yields can only be explained by a chain mechanism of transformation. To test this assumption, a study was made of the effect of the radiation dose rate on the breakdown yield of N-acetylglucosamine (Table 2.3).* It was found that the breakdown yield of N-acetylglucosamine rises with the dose rate,

---

* The lability of glucosamine solutions must be considered in the case of low dose rates, at which a prolonged exposure is necessary. Solutions of N-acetylglucosamine were therefore used to establish these relations.

which proves that the transformation of glucosamine and N-acetylglucos-amine indeed proceeds by a chain reaction. The primary products formed by the chain process have not yet been identified among the complex mixture of compounds produced by the irradiation of glucosamine solutions. However, the very fact that carbohydrate derivatives undergo transformations by a chain mechanism is pointed out here for the first time, and this finding may contribute toward a better understanding of the effects of radiations on biological systems containing glucosamine.

TABLE 2.3. Effect of dose rate and concentration on the breakdown yield of N-acetylglucosamine [8]

Concen-tration, M	Dose rate, rad/sec	Breakdown yield	Concen-tration, M	Dose rate, rad/sec	Breakdown yield
$5 \cdot 10^{-2}$	610	31	10^{-1}	563	8.8
$5 \cdot 10^{-2}$	200	7.0	10^{-1}	230	3.5
$5 \cdot 10^{-2}$	48	2.3			

It appears, therefore, that carbohydrates can be divided into two groups on the basis of their reaction to ionizing radiations. Compounds belonging to the first group (glucosamine and its derivatives) can undergo trans-formations by a chain mechanism under the influence of radiation. The transformations of compounds in the second group (monosaccharides, glycosides, disaccharides) involve an interaction with the radicals formed by the radiolysis of water (OH, H, e_{aq}) and with excited water molecules.

THE FORMATION OF MONOSACCHARIDES IN IRRADIATED AQUEOUS SOLUTIONS OF GLYCOSIDES

We analyzed the relationship between the effect of ionizing radiations and the structure of the tested compound on the basis of chemical deter-minations of the radiolytic products and the kinetics of their formation, with particular reference to processes common to the category of com-pounds considered.

As noted above, the radiolysis of all the glycosides tested yields the corresponding monosaccharide; this process is rather slow in solutions of methylglycosides but performs a major role in the case of aromatic glycosides and disaccharides (see Table 2.1). This process is better known in disaccharides and β-benzylglycoside.

In a study of the radiolysis of saccharose solutions irradiated in an oxygen atmosphere Phillips [12] concluded that this process involves an oxidative hydrolysis. It was assumed that one part of the molecule under-goes oxidation, yielding simultaneously fructose and gluconic acid or glucose and glucosone. All these compounds were indeed identified among the radiolytic products, but it was found that the yield of monosaccharides exceeds by far that of the corresponding oxidized products. This means that the monosaccharides arise largely by other pathways. It was thought that similar processes occur when irradiation is carried out in a nitrogen atmosphere (pH $= 7$). This assumption was tested by analyzing the accumulation of acids during the irradiation of 10^{-2} lactose solutions [7]. It became clear that either the acids are not primary radiolytic products at all, or they are formed as such but at a very low yield (less than 0.1). In support of this conclusion is the finding that the concentration of the acids increases during storage while that of hydrogen peroxide declines. Thus, the proposed scheme is inapplicable to this system since the formation of monosaccharides during the radiolysis of lactose solutions is not accompanied by an accumulation of acids.

TABLE 2.4. Yield of the radiolytic breakdown of 10^{-2} M solutions of disaccharides, pH $= 7$, dose range 0.9–0.8 Mrad [7, 13]

Compound	Conditions of irradiation	$G\,(-)$	G (glucose)	G (galactose)
Lactose	N_2	2.3	0.71(31)	0.43(19)
	N_2O	4.0	1.12(28)	0.6(14.5)
	N_2 pH $= 14$	3.1	1(29)	0.82(26)
Cellobiose	N_2	3.1	0.8(26)	–
	N_2O	6.5	1.5(23)	–
	O_2	2.8	1.05(38)	–
Maltose	N_2	3.3	1.0(30)	–
Gentiobiose	N_2	4.0	1.2(30)	–

Note: The figures in parentheses indicate the contribution of the given process as a percentage of the sum total of transformations.

Table 2.4 shows data on the breakdown yield of disaccharides and the formation of the corresponding monosaccharides on irradiation under different conditions. It is noteworthy that the breakdown yield from disaccharides is nearly twice as large in the presence of nitrous oxide as in a nitrogen atmosphere.

Irradiation in an N_2O atmosphere causes the transformation of e_{aq} according to the reaction

$$N_2O + e_{aq} \xrightarrow{\ H^+\ } N_2 + OH. \qquad (2.3)$$

It can be assumed, therefore, that the transformation of disaccharides involves mainly OH radicals. In other words, the reductive component of radiolysis, notably the hydrated electron, plays a minor role in the breakdown of disaccharides. Indeed, it was found [11] that the rate constant of the reaction

$$RH + OH \rightarrow R\cdot + H_2O,$$

where RH is the molecule of the polyhydroxyl-containing compound, ranges from 10^9 to 10^{10} $M^{-1}\cdot sec^{-1}$ (in the case of lactose, $k_{OH} = 3.3 \cdot 10^9$).

This means that at a solute concentration of 10^{-2} M and a dose rate of 600 rad/sec the stationary concentration of OH radicals is negligible and practically all of these radicals are utilized according to reaction (2.1). In this case the reaction $R \cdot + OH \rightarrow ROH$ is practically ruled out (although it can occur in much more dilute solutions), and stable radiolytic products are formed according to the reactions

$$R \cdot + R \cdot \rightarrow R - R; \qquad (2.4)$$

$$R \cdot + R \cdot \rightarrow A_1 + A_2, \qquad (2.5)$$

where $R-R$, A_1 and A_2 are stable products.

A similar reaction can be expected from the H atom, which also shows a fairly high rate constant with respect to many organic compounds. The hydrated electron behaves otherwise. Its reaction with the sugar

$$M + e_{aq} \rightarrow [M]^- \rightarrow R \cdot + A^- \qquad (2.6)$$

(M — carbohydrate molecule; A^- negative ion, e. g., OH^- or RO^-) is characterized by a relatively low rate constant which does not exceed 10^7 M^{-1} sec^{-1} for the carbohydrates tested ($k_{e_{aq}+glucose} \leqslant 10^6 M^{-1} sec^{-1}$) [14].

On the other hand, the rate constant of the reaction between e_{aq} and carbonyl compounds is much higher, ranging from 10^8 to 10^9 M^{-1} sec^{-1} [15]. As will be shown later, the irradiation of aqueous solutions of carbohydrates yields precisely carbonyl compounds. It can be expected, therefore, that at doses as low as 1 Mrad a considerable part of the hydrated electrons will participate in the transformation of radiolytic products and that this participation will increase in parallel with the dose. Finally, at a sufficiently high dose, varying from one compound to another, the hydrated electron no longer plays any appreciable role in the transformation of the original sugar. These factors must be considered whenever the transformation of the carbohydrate (disaccharides, for example) involves in effect solely the oxidative component of the radiolysis.

Returning to the radiolysis of disaccharides, we must consider the laws governing the formation of monosaccharides. It is evident from Table 2.4 that monosaccharides are the main radiolytic products since their yield is quite high. The yield of monosaccharides obtained by irradiation in an atmosphere of nitrous oxide is much greater than that achieved when irradiation is carried out in a nitrogen atmosphere. This means that most of the monosaccharides are formed with the participation of the OH radical. It is even more instructive to compare the yield of monosaccharides with the sum total of the products, i. e., to examine the ratio between G (monosaccharide) and G (-disaccharide). Four disaccharides were analyzed, three of which were composed entirely of glucose, while the fourth — β-1, 4-galactosylglucose — contained lactose. When these disaccharides were irradiated in a nitrogen atmosphere, glucose accounted for 26—31% of the total amount of products, i. e., the yield of glucose was practically unrelated to the structure of the disaccharide. On the other hand, the irradiation of lactose yielded additionally 19% of galactose, thus bringing the total yield of monosaccharides to 50% of the sum total of products. Similar results were obtained on irradiating these disaccharides in a nitrous oxide atmosphere, though the yield of glucose formed by the radiolysis of lactose and cellobiose is lower by 3%, and of galactose — by 4.5%. The 3–4% increase in the yield of monosaccharide as a proportion of the total amount of radiolytic products when irradiation is performed in a nitrogen atmosphere can be attributed to an involvement of the hydrated electron in the process; in other words, the contribution of the hydrated electron in the formation of monosaccharides, at the doses used, does not exceed 3–4% of the sum total of the products.

The difference in the radiolytic yield of monosaccharides from lactose and other disaccharides can be explained as follows. In lactose, glucose is a component of the aglycon moiety, and the quantity of glucose formed as a proportion of the sum total of products is practically equal to the yield of glucose from other sugars. This suggests that the glucose formed by the radiolysis of all disaccharides originates solely from the aglycon. The reaction of the OH radical with the "sugar" moiety of the molecule triggers uniform changes which lead to cleavage of the glycoside bond and formation of glucose from the aglycon. Differences in the stereochemistry or the nature of the glycoside bond apparently play only a minor role in such cases. In contrast to this, the reaction between the OH radical and the aglycon moiety of the disaccharide damages the molecule in such a manner that the radiolysis of lactose yields galactose; the irradiation of the other disaccharides does not yield monosaccharides from the "sugar" moiety of the molecule.

Irradiation of cellobiose in an oxygen atmosphere yields a greater proportion of glucose within the sum total of the products — 38% as compared with 26% in a nitrogen atmosphere. This can be explained in various ways. One possibility is that oxygen reacts with the nascent radicals to form hydroperoxide radicals

$$R\cdot + O_2 \to RO_2\cdot, \tag{2.7}$$

the breakdown of which causes the destruction of one part of the molecule and finally leads to the formation of a monosaccharide from the other part of the molecule (constituting 12% of the total amount of products). Alternatively, the disaccharide molecule may react with the O_2^- ion-radical (or with e_{aq} during the early stages of the process if irradiation is carried out in the absence of oxygen) according to the scheme

$$M + O_2^- \to [M]^- \to R\cdot + A^-. \tag{2.8}$$

The formation of monosaccharides on irradiation of cellobiose can be explained by the scheme on p. 59. This scheme, however, requires further verification.

Studies of the transformations of aromatic glucosides show that this process involves a different mechanism. It is known [16] that OH radicals react at a higher rate with aromatic compounds $(k_{OH} \approx 10^9 - 10^{10}\,M^{-1}\,sec^{-1})$. However, the rate constants of the reaction of e_{aq} with benzyl- and p-nitrophenylglucosides are also quite high [16]: $7\cdot 10^7$ and $(1 - 5)\cdot 10^9\,M^{-1}\,sec^{-1}$. This means that the oxidative and reductive components of radiolysis participate to practically the same extent in the transformations of aromatic glycosides.

Determinations of the radiochemical stability of β-benzylglucoside (Table 2.5) have shown that this compound reacts fairly readily with radicals formed from ionized and excited water molecules. On irradiation in an N_2O atmosphere, for example, the breakdown yield of β-benzylglucoside reaches 7 molecules per 100 eV of absorbed energy, which indicates a practically total capture of the OH radicals and capture of some of the radicals originating from excited water molecules. The radiolytic products of β-benzylglucoside include dibenzyl and glucose, formed at the yields shown in Table 2.5.

TABLE 2.5. The breakdown yield and yield of product formation in the radiolysis of β-benzyl-glucoside solutions, pH = 7 [9]

Concentration of β-benzyl-glucoside, 10^{-2} M	Conditions of ir-radiation	G (—)	G (glucose)	G (dibenzyl)
1	N_2	3.2	1.0(31)	0.15(4.7)
5	N_2	4.4	1.8(41)	0.22(5)
5	N_2O	7.0	0.6(8.6)	0
5	O_2	5.5	1.5(27)	0.08(1.5)

Note. The figures in parentheses indicate the contribution of the given process as a percentage of the sum total of transformations.

It was found that the yield of dibenzyl rises slowly with the concentration, reaching 0.12, 0.15, 0.22 and 0.24 molecules per 100 eV at concentrations of $5 \cdot 10^{-3}$, 10^{-2}, $5 \cdot 10^{-2}$ and 10^{-1} M, respectively. This suggests that excited water

molecules do not participate in the formation of dibenzyl, since otherwise
its yield would increase sharply with the concentration. The yield of dibenzyl is influenced by the presence of various compounds serving as acceptors of the hydrated electron. Thus, the yield of dibenzyl is practically
nil if irradiation is carried out in an acidic medium (at pH values of 3, 7,
and 12 the respective yields are 0.04, 0.22, and 0.17) or in the presence of
nitrous oxide. The data given in Table 2.5 show that dibenzyl arises solely
with the participation of e_{aq}. It is noteworthy that dibenzyl also appears when
β-benzylglucoside is irradiated in an oxygen atmosphere; the O_2^- ion presumably functions in a similar way, transmitting its charge to the benzylglucoside molecule.

The process leading to the formation of glucose is more complex. Part
of the glucose certainly appears simultaneously with the dibenzyl, as may
be seen by comparing the yields of these compounds in nitrogen and oxygen
atmospheres. Thus, the yields of dibenzyl and glucose in an oxygen atmosphere are lower by 0.14 and 0.3, respectively, which means that two glucose
molecules appear for each dibenzyl molecule formed. This is not surprising since dibenzyl can arise only by the recombination of benzyl radicals.
Part of the glucose is evidently formed by the interaction of benzylglucoside
with the OH radical, since the yield of glucose in a nitrous oxide atmosphere
is 0.6. In a nitrogen atmosphere the yield of glucose must be only half the
value, i.e., 0.3.

Thus, the yield of glucose (1.8) formed as a result of irradiation of β-
benzylglucoside in a nitrogen atmosphere includes 0.3 molecules/100 eV
contributed by e_{aq} and another 0.3 molecules/100 eV resulting from the
participation of the OH radical. However, the bulk of the glucose yield —
1.2 molecules — remains unexplained. Possibly this amount arises from
transformations of the dihydroaromatic compounds formed by disproportionation reactions.

The transformations resulting from the irradiation of β-benzylglucoside
solutions can therefore be outlined as follows:

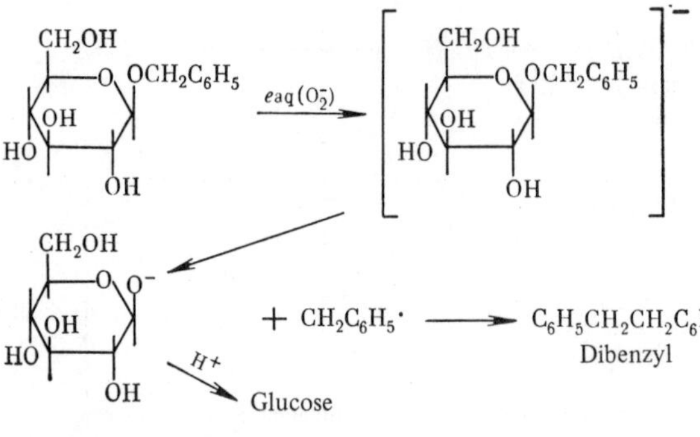

The interaction of benzylglucoside with the hydrated electron yields a
negatively charged molecular ion which then breaks down into a glucose ion
and a benzyl radical.

The recombination of benzyl radicals yields dibenzyl. Indeed, the spectra obtained by the pulse radiolysis of β-benzylglucoside solutions show a bend at 315 nm against the background of the main peak (320 nm on irradiation in an He atmosphere). This bend is totally absent when β-benzylglucoside solutions are irradiated in a N_2O atmosphere. Comparison of the two spectra after normalization reveals a differential spectrum corresponding [9] to the benzyl radical. These findings, together with the results of irradiation in a stationary regime, fully confirm the pertinent part of the scheme.

The irradiation of β-benzylglucoside solutions in an oxygen atmosphere similarly results in an absorption corresponding to the benzyl radical but in this case it has a much lower intensity. This constitutes further proof of the above conclusion that the O_2^- ion-radical can react in the same way as e_{aq}. However, the main absorption observed on pulse irradiation has a peak at 320 nm. The absorption intensity is practically the same irrespective of whether the samples are irradiated in He, N_2O, or O_2 atmospheres. This means that the absorbing compound is formed with the participation of OH, H, e_{aq} and O_2^- radicals, and that in all cases the reaction products have practically the same molar extinction coefficients and consequently contain identical chromophore groups.

Neta et al. attribute the absorption at 320 nm in their experiments to a cyclohexadienyl radical [17]. Similar structures may also be formed in the above case. Indeed, the reaction between the aglycon and the OH radical must yield an addition product with the structure of an oxycyclohexadienyl radical. Such reactions are well known in the radiochemistry of aqueous solutions of aromatic compounds. A structurally similar radical arises by the action of the H atom, as well as by that of e_{aq} and O_2^-, as a secondary (but major) reaction product:

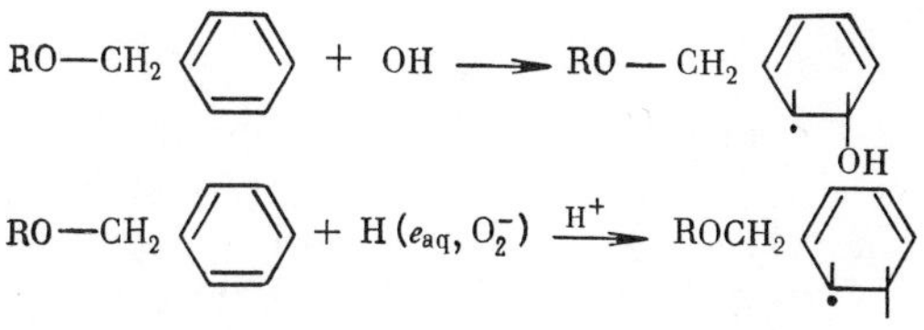

Other addition reactions are also possible. Since the stable products are formed as a result of disproportionation reactions, the following structures may be expected:

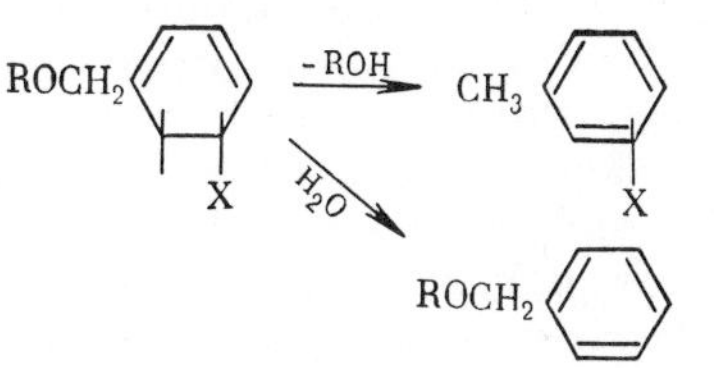

where RO is the glucose residue

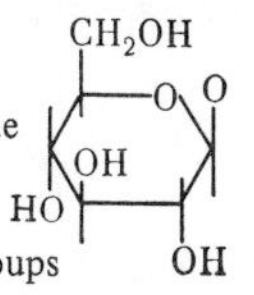

and X represents H or OH groups

The breakdown of the dihydroaromatic compounds may serve as an additional source of glucose.

The presence of phenols in the irradiated solutions prove that such transformations can take place. These or similar processes may serve as additional sources of glucose.

An essentially similar scheme for the transformations of aromatic glucosides was proposed by Phillips [19] in a study of the transformations of phenylglucoside. Among the radiolytic products obtained by the pulse irradiation of a phenylglucoside solution Phillips detected the phenoxyl radical $C_6H_5O \cdot$, the formation of which he explains as follows:

where R is the glucose residue

The cyclohexadienyl radical I formed by addition of an OH radical to the aromatic ring loses a hydroxyl ion and turns into a positively charged molecular ion II, the breakdown of which yields the phenoxyl radical IV and the glucose ion III. Radical IV reacts with phenylglucoside to form phenol, while the glucose ion undergoes further transformation according to the scheme below

The following conclusions can be drawn as regards the formation of monosaccharides from glycosides generally.

In the radiolysis of glycoside solutions, the bulk of the monosaccharides is formed as a result of the decomposition of one moiety of the molecule. The transformations of disaccharides are brought about mainly by the action of the OH radical, whereas those of benzylglucoside involve also the reductive component of radiolysis, namely, H and e_{aq}. In the radiolysis of benzylglucoside, part of the glucose is formed as a result of the dissociative addition of e_{aq}; similar processes (involving the breakdown of a negatively charged molecular ion) possibly take place in other cases as well.

THE FORMATION OF DEOXYSACCHARIDES IN THE IRRADIATION OF AQUEOUS SOLUTIONS OF CARBOHYDRATES

The mechanism of radiochemical reactions cannot be understood without knowledge of the specific structures arising during the process. It is known [19] that the irradiation of carbohydrates under various conditions yields mainly the following categories of compounds: monosaccharides from glycosides; oxidation products retaining the original structure (carbonyl compounds and acids); oxidized fragments containing fewer carbon atoms, and polymerization products [20, 21]. However, the data available are not sufficient to explain the mechanisms involved. Attempts have been made to interpret the results obtained in terms of multistage processes (e. g., the consecutive action of two or more radicals) but so far there is little kinetic support for such hypotheses.

Studies of numerous compounds, mainly glycosides of varying structure, have shown that the radiolytic products obtained in dilute solutions in a nitrogen atmosphere — i. e., in the presence of the radicals OH, e_{aq} and H — include a large group of compounds characterized by a molecule containing a deoxyunit [22]. Thus, 5-deoxy-4-ketohexose- and methyl-6-al-5-deoxy-hexafuranoside are among the main products of the radiolysis of α-methyl-glycoside [23, 24]. According to chromatographic data, these compounds do not appear on irradiation in an oxygen atmosphere; their formation can therefore be attributed to the action of the hydrated electron. The possible sequence of reactions is shown on pp. 63—64 (the stereochemistry of the original sugar is retained here and in the following formulae for the sake of convenience, although this has not been established in the majority of cases).

The methylglycoside molecule reacts with e_{aq} to form the negatively charged ion I, which undergoes an intramolecular breakdown to yield the ion-radical II. The latter disproportionates into the ions III and IV which differ from one another in the position of the double bond (only the oxidation products are shown in the scheme). Ion IV forms methyl-6-al-5-deoxyhexa-furanoside (V) by an intramolecular nucleophilic substitution. Such transformations may explain the formation of a number of other products which appear in the radiolysis of glycosides.

On the other hand, if a water molecule or ROH is detached from the intermediary radicals, the formation of analogous deoxycompounds and of the preceding compounds, with the participation of radicals from the radiolysis of water (OH and H), may be outlined as follows:

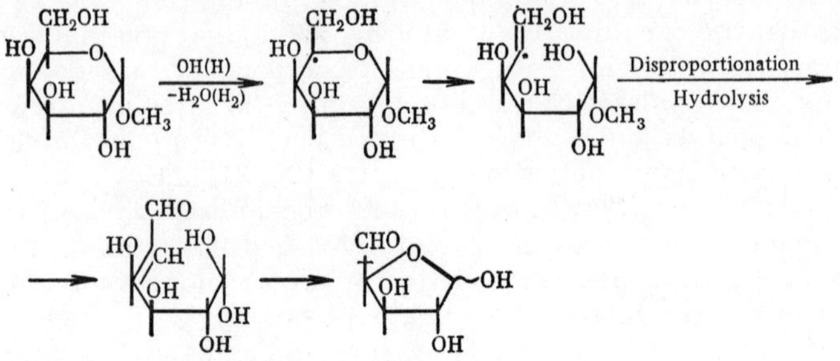

Two isomeric compounds with the structure of 2-deoxy-4-ketoderivatives of methylgalactoside have also been obtained from methylgalactoside samples As in the former case, the formation of these compounds may be attributed to the action of OH radicals:

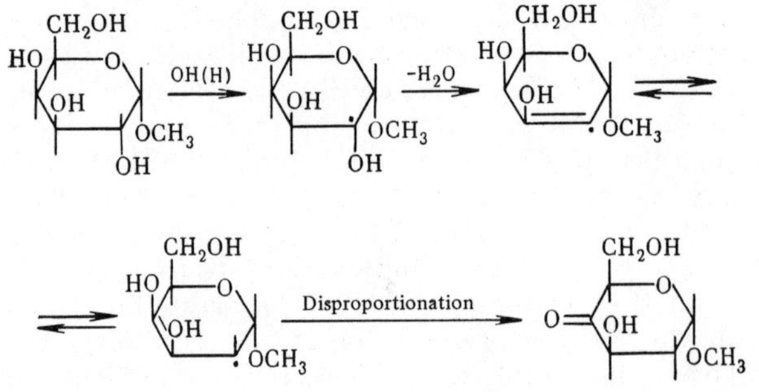

A similar picture is observed in the radiolysis of other compounds. Thus, the irradiation of β-methylarabinose yields lactones of 4-deoxy-pentonic acid and 4-deoxy-3-ketopentonic acid, which are possibly formed with the participation of OH (H) radicals:

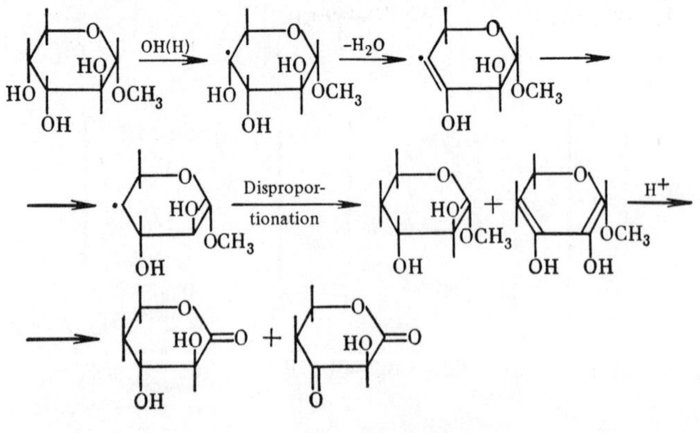

It is noteworthy that the process yields mainly compounds which can be regarded as the two products of a disproportionation reaction. Thus, the radiolysis of cellobiose yields two compounds of the deoxysaccharide type, namely, 4-deoxyhexose (IX) and 2-deoxy-3-ketohexose (X), the formation of which can be attributed not only to the action of the hydrated electron:

but also to the OH (H) radical:

Pathway *a* was discussed above; pathway *b* comprises a new element — migration of the free valence by a single-electron shift and subsequent rupture of the −C−O− bond.

The radiolysis of cellobiose and lactose yields, among other products, two pentoses — arabinose and lyxose. The presence of such products indicates that the −C−C− bond in the "sugar" moiety of the disaccharide molecule is cleared in both cases. These compounds probably arise as a result of reactions involving the OH radical

It is assumed in this case that cleavage of the C—C bond takes place after the migration of the free valence, similarly brought about by a single-electron shift.

One of the products obtained by the irradiation of α-methylglycoside solutions is α-methyl-6-al-glycopyranoside [25]. This compound appears upon irradiation in nitrogen and oxygen atmospheres, i. e., its formation is associated with the action of the OH radical:

The proposed scheme for the formation of the products is not free from contradictions since it is based solely on the structure of the initial and final products. A discussion of the principles involved in the proposed scheme is nevertheless worthwhile. One of the contradictions inherent in the proposed schemes is that the formation of practically all the products may be attributed to the action of the ion-radical e_{aq} as well as to that of the radicals OH and H. Chromatography shows that two compounds obtained by the irradiation of α-methylglycoside are formed by the action of the hydrated electron. On the other hand, chromatography reveals identical products in the radiolysis of lactose solutions in N_2, N_2O and O_2 atmospheres. Moreover, the composition of the products may change in the course of isolation (during storage, enrichment, chromatographic fractionation, etc.), so that the compound isolated is not necessarily a major product of the radiolysis even if it predominates in the enriched mixture.

Thus, the structure of a radiolytic product does not, as a rule, indicate unequivocally the way in which it was formed.

The proposed schemes are based on the following assumptions.

1. The final products of radiolysis are formed solely by the disproportionation of radicals.

2. The double bond of carbohydrate radicals or stable products can migrate by means of an enediol tautomerism or allyl rearrangement.

Such cases are widely known in carbohydrate chemistry [26]. This question, however, must be reconsidered in each individual case according to the specific structure of the reacting compound.

3. In the reaction between the OH radical and pyranosides, the free valence is located mainly at the C_1 and C_5 positions, as demonstrated by EPR determinations made on frozen aqueous solutions of carbohydrates [10]. However, localization of the free valence at other positions cannot be ruled out.

4. The reaction with the hydrated electron yields a negatively charged molecular ion. This is evident from EPR data on frozen aqueous solutions of carbohydrates [27].

5. The negatively charged molecular ion breaks down into a negative ion and a radical. Analogies are known from the literature [28].

6. The free valence can migrate as a result of the single-electron shift accompanying cleavage of the C—O and C—C bonds. EPR data on frozen aqueous solutions of many compounds confirm the possibility of such processes [27].

7. The radicals formed can split off elements of water and alcohol.

Thus, practically all the basic assumptions in the above schemes for the transformation of carbohydrates on irradiation of their aqueous solutions are confirmed to a varying extent by experimental evidence.

In order to determine the origin of the deoxycompounds, a study was made of their accumulation as a function of the conditions of irradiation and the object irradiated. The formation of deoxysugars is to some extent linked with a dehydration or detachment of an ROH (alkoxyl) group. EPR studies [29] of frozen aqueous solutions of inositol have shown that the irradiation of aqueous solutions of this compound yields radicals resulting from cleavage of the C—H bond, apparently by a reaction with the OH radical. With a subsequent rise in temperature the original radicals spontaneously split off water and turn into radicals containing the ketodeoxy group

A similar transformation was observed earlier in aqueous solutions by the EPR method (using the flow technique and solutions of sulfuric acid and titanium salts), though under considerably more rigid conditions [30]. Thus, one of the possible transformations of the radicals originating from polyhydroxyl compounds involves a dehydration, ultimately leading to the formation of deoxycompounds. This process was studied quantitatively by tracing the accumulation of malondialdehyde, which is formed immediately in the solution as a result of the irradiation; the samples examined were irradiated solutions oxidized with periodic acid and solutions reduced with potassium borohydrate, irradiated, and then oxidized with periodic acid.

Table 2.6 contains data on the yield of malondialdehyde immediately after irradiation under different conditions. Malondialdehyde appears in practically all the samples examined, though at a very low yield, measured in hundredths of a molecule per 100 eV of absorbed energy. It is of interest that the irradiation of inositol yields practically no malondialdehyde. This can be explained on the basis that the formation of malondialdehyde necessarily involves the rupture of a C−C bond. One such rupture is sufficient for carbohydrates but two are necessary in the case of inositol — naturally a much less probable event. Comparison of the results obtained shows that malondialdehyde is formed with the participation of the OH radical, since the yield of the latter increases on irradiation in an N_2O atmosphere. The increased yield of malondialdehyde in an acidic medium may be attributed to the hydrolytic breakdown of certain unstable products [31]. The following scheme of the formation of malondialdehyde may be proposed on the basis of the results obtained:

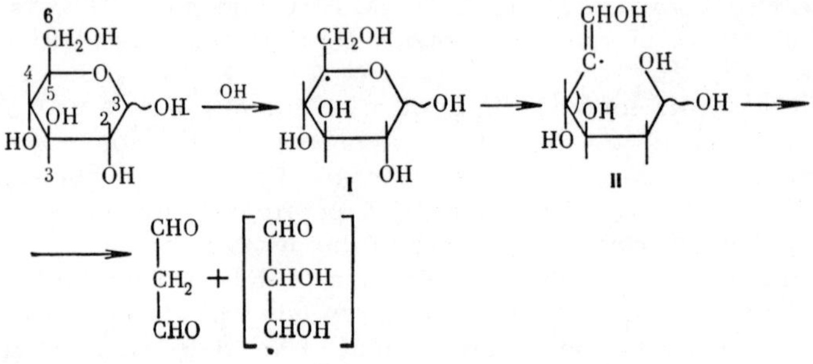

The settling of the free valence at C_5 (I) is followed by opening of the pyranose ring with the formation of the radical II. One of the possible transformation pathways of the latter involves cleavage of the C_3-C_4 bond as a result of a single-electron shift, which yields malondialdehyde as one of the products.

TABLE 2.6. Yield of malondialdehyde following irradiation of carbohydrate solutions

Compound, 10^{-2} M	Conditions of irradiation			
	N_2	N_2O	O_2	N_2 (pH= 2)
α-Methyl-D-galactoside	0.01	0.03	0.01	0.03
β-Methyl-D-mannoside	0.02	0.04	0.01	0.01
α-Methyl-D-glycoside	0.000	0.02	0.01	0.01
Galactose	0.02	0.04	−	0.03
Mannose	0.01	0.06	−	0.04
Glucose	0.02	0.04	−	0.04
Inositol	0.00	0.00	0.00	0.00

According to the above scheme, malondialdehyde is formed from the nonreducing part of the monosaccharide. This explains the observed similarity in the behavior of monosaccharides and glycosides.

TABLE 2.7.   Yield of deoxysugars in irradiated solutions of carbohydrates

Compound, 10^{-2} M	Conditions of irradiation			
	N_2	N_2O	O_2	N_2 (pH = 2)
α-Methyl-D-galactoside	0.27	0.27	0.08	0.12
α-Methyl-D-mannoside	0.22	0	0.23	0
α-Methyl-D-glycoside	0.22	0.06	0	0
D-Galactose	0.10	0.15	–	0.13
D-Glucose	0.13	0.13	–	0.05
D-mannose	0.10	0.22	–	0.06
Lactose	0.22	0.22	0.22	–
Cellobiose	0.22	0.22	0.10	–
Starch (0.07% solution)	0.22	–	–	–
Blood group factor (0.1% solution)	0.22	0	–	–
Inositol	0.08	0	0	0

Table 2.7 shows the yield of malondialdehyde (deoxysugar) under various conditions immediately after the oxidation of the irradiated solutions with periodic acid. The oxidation of carbohydrates obviously yields malondialdehyde in those cases in which $-CHOH-CH_2-CHOH-$ or $CHO-CH_2-CHOH-$ fragments are present together with hydroxyl or carbonyl groups ($-CHOH-$ and $-CO-$).

It is evident from Table 2.7 that the yield of deoxysugars varies greatly with the conditions of irradiation (from 0 to 0.24). This relationship is most pronounced in the case of irradiation of inositol solutions. Indeed, the formation of malondialdehyde after oxidation with periodic acid was detected in only one case – in neutral solutions irradiated in a nitrogen atmosphere. On these grounds the formation of deoxycompounds can be attributed entirely to a reaction with the hydrated electron.

Such a scheme has been discussed previously on several occasions. A similar explanation probably applies to the formation of deoxycompounds in the radiolysis of methylmannoside (with the participation of e_{aq} or O_2^-) since the yield of deoxycompounds drops to zero in the case of irradiation of acidic (pH = 2) or N_2O-saturated solutions. Such a clear-cut relationship does not exist in the other cases.

According to the above scheme the deoxycompounds must appear simultaneously with deoxyketocompounds in which the carbonyl group and the deoxy unit are adjacent. If the carbonyl group occupies a terminal position ($CHO-CH_2-$), such compounds appear as the oxidation products of disproportionation reactions and are determined by the above method as reduction products, that is, as true deoxycompounds. In other words, after reduction with potassium borohydride these compounds will not yield malondialdehyde on subsequent oxidation. This appears to be the case with the radiolysis of methylgalactoside and methylmannoside solutions, in which the malondialdehyde content decreases after reduction. However, compounds of this type can also appear as a result of other processes (with the participation of the OH radical, for example).

The radiolysis of solutions of α-methylglycoside does not yield any appreciable amount of malondialdehyde. As noted above, the radiolytic products of α-methylglycoside include compounds with the deoxy unit at position 5. The oxidation of these compounds yields β-oxypropylonic aldehyde or the corresponding acid. The yield of deoxysugars in the radiolysis of α-methylmannoside solutions is the same (0.23) whether the samples are irradiated in nitrogen or oxygen atmospheres. After reduction and subsequent oxidation, however, malondialdehyde was not detected in these systems. It can be assumed on these grounds that both processes lead to structures of the same type, containing ($-CH_2-CHO$) fragments, as a result of the action of both e_{aq} and O_2^-.

Such structures are totally absent when the irradiation is carried out in an N_2O atmosphere or at pH = 2; in these cases the negative ion-radicals are evidently transformed into products with a different structure.

The results obtained indicate that e_{aq} or the O_2^- ion react with the carbohydrate molecule according to the following scheme:

The end products are deoxy- (II) and deoxyketo- (III) compounds. The latter are determined by the method described as deoxysugars if their carbonyl group occupies a terminal position. Such compounds may also be formed if the compound of type II already contains an aldehyde group. Deoxyketocompounds of type III (Table 2.8), after reduction with borohydride and subsequent treatment with periodic acid, can also yield malondialdehyde under certain conditions, and this property can be used for the quantitative determination of such compounds.

TABLE 2.8. Yield of deoxyketo sugars in irradiated solutions of carbohydrates

Compound	Conditions of irradiation			
	N_2	N_2O	O_2	N_2 (pH = 2)
α-Methyl-D-glactoside	0.06	0.12	0.04	0.1
α-Methyl-D-mannoside	0.09	0.22	0.06	0.43
α-Methyl-D-glycoside	0	0	0	0
D-Galactose	0.22	0.40	−	0.40
D-Glucose	0.17	0.25	−	0.29
D-Mannose	0.15	0.25	−	0.17
Lactose	0.42	0.44	0.44	−
Cellobiose	0.42	0.63	0.22	−
Starch (0.07% solution)	0.23	−	−	−
Blood group factor (0.1% solution)	0.22	−	−	−
Inositol	0.38	0.42	0.37	0.45

By summarizing the data available on the formation of deoxycompounds, the following conclusions may be drawn.

The ion-radical e_{aq} (or O_2^- in some cases) causes detachment of the OH group and thus yields compounds containing $-CH_2-$ and $-CH_2-\overset{\|}{\underset{O}{C}}-$ units.

The radicals OH and H (possibly, HO_2) break the C−H bond. The radicals formed in the process undergo dehydration (or loss of the ROH group) and disproportionation and form compounds containing the $-CH_2-\overset{\|}{\underset{O}{C}}-$ unit.

The position of the carbonyl group and deoxy unit depends on the specific properties of the molecules formed. These groups can migrate by means of ketoenol tautomerism and allyl rearrangement, which may lead to the formation of identical compounds by different pathways.

The above processes occur at relatively low dose rates, when reactions such as

$$R \cdot + OH \rightarrow ROH;$$

$$R \cdot + e_{aq} \xrightarrow{H^+} RH$$

are practically impossible. At the same time, these reactions may occur when biological systems are exposed to radiation and perform an important function in the mechanism governing the biological action of ionizing radiation.

THE TRANSFORMATION OF MONOSACCHARIDES IN FROZEN AQUEOUS SOLUTIONS

The irradiation of frozen aqueous solutions of pentoses indeed triggers processes involving stereochemical changes which can be explained by the reactions mentioned above. Frozen aqueous solutions of several pentoses and hexoses were examined. Under the conditions of the experiment $(10^{-2}$ M, frozen polycrystalline samples, -78°C), EPR measurements [32] detect only radicals formed from the irradiated sugar and not radicals resulting from the radiolysis of water (OH, e), this being apparently due to the low mobility of the solvates of radicals arising from the sugar and the great mobility (and reactivity) of the radicals originating from the radiolysis of water.

Chromatographic analysis of D-ribose solutions irradiated in a nitrogen atmosphere has shown that D-arabinose, lyxose and xylose are among the major products of the radiolysis. By a similar procedure lyxose was isolated from frozen samples of D-arabinose, and lyxose and arabinose, respectively, from D-xylose and D-lyxose. Isomeric pentoses other than those mentioned above were not detected in the irradiated preparations. Similar processes probably take place as a result of the irradiation of hexoses, though here they follow a more complex course. Thus, ribose was identified chromatographically in irradiated solutions of glucose; the radiolysis of galactose yields glucose and xylose, and that of mannose — ribose and xylose.

Epimerization was shown to be associated with the indirect effect of radiation since it does not take place during the irradiation of solid ribose. The epimerization process takes place in frozen aqueous solutions at temperatures ranging from -78 to 0°C but not in a liquid phase or at -196°C. This can be attributed in the former case to the low stationary concentration of the radicals, and in the latter case — to the practical immobility of the radicals formed. The participation of radicals originating from the radiolysis of water in the epimerization process is evident from the following data. Irradiation of frozen ribose at pH = 2 in a nitrogen atmosphere sharply alters the qualitative composition of the products. Arabinose is the only product in this case; no trace of lyxose or xylose appears. The same result is obtained when the samples are irradiated in a N_2O atmosphere. It is evident from these findings that the stabilized electron participates in the epimerization process. The latter process is clearly associated with the action of e. Thus, the formation of arabinose from ribose can be outlined as follows:

To sum up, the irradiation of aqueous solutions of carbohydrates triggers a wide variety of transformations of these compounds. Generally speaking, when secondary reactions with the radiolytic products of water of the type

$$R\cdot + OH \rightarrow ROH$$

are inhibited, the radicals formed are either disproportionate, yielding oxidized products of the carbohydrate, or turn into deoxyketocompounds by losing water.

CONCLUSION

The transformation of carbohydrates in dilute aqueous solutions under the influence of radiation constitutes a complex multistage process which depends not only on the conditions of irradiation but also on the structure of the particular sugar. In particular, the final stages of the process and the specific structure of the radiolytic products depend largely on the structure of the carbohydrate.

The damage to carbohydrates results mainly from the action of the OH radical. The hydrated electron performs a comparatively minor role during the initial stages of the process. For modern radiobiology the most urgent task is to determine the structure of the molecules or molecular fragments modified by radiation, since knowledge of the actual chemical changes in the structure of biopolymers will provide a basis for further research aimed at the elimination of radiation-induced changes by chemical or biochemical means.

BIBLIOGRAPHY

1. Phillips, G.O.— Nature 173 (1954).
2. Bothner, C.T. and E.A.Balazs.— Radiat. Res. 6 (1957), 302.
3. Wolfrom, M.L. et al.— Abstr. Pap. Am. Chem. Soc. 130 (1956), 16A.
4. Kochetkov, N.K. et al.— Zh. Obshch. Khim. 35 (1965), 268.

5. Kochetkov,N.K. et al.— Zh. Obshch. Khim. 35 (1965),2246.

6. Kudryashov,L.I. et al.— Zh. Obshch. Khim. 40 (1970),1133.

7. Kochetkov,N.K. et al.— Zh. Obshch. Khim. 36 (1966),229.

8. Kudryashov,L.I. et al.— Zh. Obshch. Khim. 38 (1968),2380.

9. Fel',N.S. et al.— Khimiya Vysokikh Energii 5 (1970),241.

10. Kochetkov,N.K. et al.— Khimiya Vysokikh Energii 2 (1968),556.

11. Scholes,J. et al. Pulse Radiolysis.— L.—N.Y., Academic Press (1965),151.

12. Phillips,G.O. and C.J.Moody.— J. Chem. Soc. 155,1960.

13. Kudryashov,L.I. et al.— Zh. Obshch. Khim. 1 (1971),41.

14. Davies,J.V. et al. Pulse Radiolysis.— L.—N.Y., Academic Press (1965),181.

15. Anbar,M. and P.Neta.— Int. J. Appl. Radiat. Isotopes 16 (1967),493.

16. Phillips,G.O. et al.— Carb. Res. 16 (1971),89.

17. Neta,P. and L.M.Dorfman.— Adv. Chem. Ser.,81 Am. Chem. Soc. (1968),222.

18. Phillips,G.O. et al.— Carb. Res. 16 (1971),105.

19. Phillips,G.O.— Adv. Carbohyd. Chem. 16 (1961),13.

20. Barker,S.A. et al.— J. Chem. Soc. (1959),2648.

21. Starodubtsev,S.V. and V.V.Generalova.— Izv. AN UzbSSR,phys. ser. 7 (1963),46.

22. Kochetkov,N.K. et al.— Izv. AN SSSR,chem. ser., No.146 (1970),201.

23. Kochetkov,N.K. et al.— Zh. Obshch. Khim. 38 (1968),79.

24. Kochetkov,N.K. et al.— Doklady AN SSSR 179 (1968),1385.

25. Kudryashov,L.I. et al.— Izv. AN SSSR,chem. ser. 189,1969.

26. Kochetkov,N.K. et al.— Carbohydrate Chemistry.— Moscow, "Khimiya," 1967. (Russian)

27. Sharpatyi,V,A. et al.— Khimiya Vysokikh Energii 2 (1968),286.

28. Pikaev,A.K. The Solvated Electron in Radiation Chemistry.— Moscow, "Nauka." 1969.

29. Nikitin,I.V. et al.— Doklady AN SSSR 190 (1970),635.

30. Norman,R.O.C. and R.J.Pritchett.— J. Chem. Soc. B (1967),1329.

31. Kochetkov,N.K. et al.— Zh. Obshch. Khim. 41 (1971),449.

32. Kochetkov,N.K. et al.— Doklady AN SSSR 183 (1968),376.

Chapter 3

THE STUDY OF RADIATION-INDUCED TRANS-
FORMATIONS IN PROTEIN SOLUTIONS BY THE
CHEMILUMINESCENCE METHOD

INTRODUCTION

The effects of the chemiluminescence* induced by various kinds of radiation are of great interest. One of the current objectives in radiobiological
research is to gain a more thorough understanding of such processes and
obtain quantitative data on all the chemical reactions and elementary
physical events taking place from the moment of exposure to γ-quanta or
light to the appearance of a weak glow.

At the moment the processes of induced chemiluminescence in comparatively complex objects (proteins, peptides, chlorophyll) are better known
than those occurring in simpler compounds. The reason for this is probably
that radiation effects in simpler systems can be studied by a number of
direct methods yielding unequivocal data; the radiolysis and photolysis of
simple systems can be analyzed at high radiation doses, and all the pathways
involved in the transformation of the tested substance can be traced. In
the case of proteins and other complex natural products exposed to large
doses of radiation, on the other hand, the conventional physicochemical
procedures can as a rule only detect secondary changes caused by chemical
reactions of amino acid residues or by alterations in the conformation of
the macromolecule. The study of the primary changes in proteins therefore
requires highly sensitive methods capable of detecting even minor changes
induced by small radiation doses. Similar considerations apply also to
other complex biological systems.

Apart from its purely theoretical value, induced chemiluminescence can
play an important part in the solution of numerous problems in radiation
chemistry, photochemistry, photobiology, and radiobiology.

REVIEW OF WORKS ON RADIATION-INDUCED
CHEMILUMINESCENCE

Research into radiation-induced chemiluminescence began with the
studies of Ahnstrom et al. and Strehler et al. [1, 2]. The authors of one

---

* The terms "chemiluminescence," "glow," and "afterglow" are used synonymously in this chapter.—
  The Editor.

of these publications discovered photosynthetic chemiluminescence, while the other work describes the glow observed following the solution of various previously irradiated substances. By now the literature contains more than 100 publications concerned with the phenomenon of induced chemiluminescence.

TABLE 3.1. Studies of radiation-induced chemiluminescence

Kind of radiation	Object of study	References
γ - and X rays	Solid phase and on solution of previously irradiated substances:	
	inorganic substances	[1, 3−5]
	low molecular weight organic substances	[1, 4, 5]
	proteins and synthetic polymers	[6−10]
	Solutions:	
	low molecular weight organic substances	[11−14, 16]
	amino acids and peptides	[14−16]
	proteins and synthetic polymers	[14, 15, 17]
	previously heated irradiated solutions	
	of DNA	[18−20]
	Complex biological objects	[21−24]
UV light	Solid phase − proteins	[25−29]
	Solutions:	
	low molecular weight organic substances	[30]
	water	[31−32]
	amino acids and peptides	[33−36]
	proteins	[26, 33, 37−56]
	Complex biological objects	[25, 57, 58]
Visible light (sensitized photochemiluminescence)	Solutions:	
	low molecular weight substances	[59−62]
	chlorophyll and other pigments	[63−67]
	amino acids and peptides	[68, 69]
	proteins	[45, 68, 38]
	Complex biological objects	[29, 70−75]

These studies can be divided into several groups according to the kind of radiation used: ionizing radiation, UV, or visible light. Each group may be subdivided into works dealing with the glow observed in the solid phase or on solution, in previously irradiated solutions or on heating previously irradiated solutions (Table 3.1). This chapter deals mainly with the chemiluminescence of protein solutions. A prolonged "phosphorescence" of protein solutions at room temperature was discovered in 1958 [37]; this finding passed unnoticed, however, and a new report appeared in 1962, describing a prolonged afterglow following the UV-irradiation of dry proteins and their solutions. Chemiluminescence in crystalline proteins as a result of X- and γ-irradiation was discovered in 1963 [7] and in protein solutions in 1964 [17]. Sensitized photoluminescence was also discovered in 1964 [38].

At the beginning, these effects were interpreted in a highly speculative fashion in terms of extraordinary "phosphorescence" or conservation of energy in the protein molecule, or by applying the concepts of solid-state physics to the dissolved protein macromolecules (excitons), etc. Many workers assumed that these discoveries were essentially new phenomena associated with the protein macromolecule as a whole, which vanish with the transition from proteins to low molecular weight compounds. Further research, however, caused this view to be abandoned. It was found that the afterglow of proteins in the solid phase is in fact an induced chemiluminescence resulting from an oxidative breakdown involving free radicals [6—10]. Studies of induced chemiluminescence in solutions showed that these processes are highly complex, consisting of several stages, and result from the formation and disappearance of free radicals.

Eventually it became clear [46] that the formation of free radicals is associated with the breakdown of tryptophyl residues in the protein molecule and that the chemiluminescence results from a kynurenine-oriented pathway in the photo- or radiation-induced oxidation of tryptophan [53]. These findings cleared the way for the application of induced chemiluminescence to the solution of various problems. Scientists acquired a "luminous label" signaling the breakdown of tryptophan residues in the protein molecule. This label found various applications, notably in radiobiology. In addition, chemiluminescence proved a valuable tool for the study of conformational changes in proteins. It was found that some of the parameters of photochemiluminescence are highly sensitive to changes in the tertiary structure of the protein. Although the causes of these changes are not quite clear, as yet, the sensitivity of this method exceeds that of most others [49, 51, 52, 54—56]. Attempts are under way to determine the conformational changes occurring in proteins in the cell by this method [57]. Induced chemiluminescence is being used in the study of various aspects of the photochemistry and radiation chemistry of proteins, the photodynamic effect, and the detection of changes in the structure of macromolecules.

GENERAL SCHEME OF THE PROCESSES LEADING
TO CHEMILUMINESCENCE IN IRRADIATED
SOLUTIONS

A simple qualitative scheme of the processes involved in radiation-induced chemiluminescence can be drawn up on the basis of data published in the literature and our own findings. The whole process may be subdivided arbitrarily into three stages: the radiation-induced formation of various chemically active compounds; diverse reactions of these compounds with the appearance of product molecules in an excited-electron state at one of the elementary stages; and finally, deactivation of the excited product molecules and emission of light by chemiluminescence.

The processes taking place during the first stage vary according to the type of radiation. The gamma-irradiation of aqueous solutions initially yields active products of the radiolysis of water: the positive water ion, hydrated electron, excited water molecule, hydrogen atom, hydroxyl radical, hydroperoxide radical, hydrogen peroxide, etc. These active radiolytic products of water react with the substance introduced to yield chemically active compounds. Indirect effects also predominate in dilute nonaqueous solutions. At high concentrations there is a direct ionization or excitation which similarly leads to specific chemically active products.

In the sensitized reaction to light the process begins with the formation of excited-electron sensitizer molecules which react with the compounds introduced and create unstable products of this.

UV irradiation populates the singlet and triplet levels of the molecules, which provide the energy feeding the chemical transformation. The chemically active products formed in the primary processes and essential for the manifestation of chemiluminescence vary greatly in nature. These comprise free radicals, free peroxide radicals, peroxides and hydroperoxides, endoperoxides, singlet oxygen, and other unstable products. The chemiluminescence method usually has a low time resolution (about 0.1 sec) and therefore cannot detect primary processes or shortlived states.

The second major stage includes the numerous transformations of the chemically active products and the low-probability event of the formation of a compound in an excited-electron state in one of the reactions. One of the main objectives is to detect this reaction. It is particularly important to establish all the pathways involved in the transformation of the chemically active compounds because the events leading to excitation very often lie outside the main course of the reaction.

According to the available data, the reactions raising molecules to an excited state include disproportionation of peroxide and oxide radicals, recombination of radicals, breakdown of peroxides involving the formation and disproportionation of radicals, decomposition (rearrangement) of peroxides, etc. As a rule, induced chemiluminescence requires the presence of oxygen during the irradiation. A high exothermicity of the primary event (more than 60 cal/mole) is obligatory for the formation of excited molecules of the product. As yet little is known about the nature of the chemical excitation associated with chemiluminescence or the factors determining it.

Finally, the third stage consists of deactivation of the excited product and emission of light by chemiluminescence. This stage is analogous to the processes of normal luminescence by optical excitation. As in the latter phenomenon, processes such as quenching, energy transfer, etc., are possible here. It is difficult to identify the source of the emission. Luminescence centers have never been detected in radiation-induced chemiluminescence in any system. In some cases of dark chemiluminescence it has been possible to identify the nature of the emitter and the type of electron transfer responsible for the glow. It was shown, for example, that the emission in certain liquid-phase reactions originates from a triplet-state ketone arising from the disproportionation of free peroxide radicals.

These are, briefly, the basic stages and processes responsible for induced chemiluminescence. Each particular system comprises definite reactions involving a variety of compounds; hence the need to detect each consecutive stage of the process and determine its quantitative parameters in order to express the whole process in kinetic, radiochemical and photochemical terms. No object has so far been analyzed in this way. The most detailed data available concern the photochemiluminescence of chlorophyll, certain dyes [62—67], glycyltryptophan, and several proteins.

RADIATION-INDUCED CHEMILUMINESCENCE IN SOLUTIONS OF AMINO ACIDS, PEPTIDES, PROTEINS AND OTHER COMPOUNDS

We shall begin with a practical description of the induced photoluminescence of certain compounds and the methods of research [74].

The study of the low-intensity chemiluminescence which appears in many solutions following their irradiation with UV-, X-, and gamma rays faces a number of obstacles. These radiations cause a prolonged afterglow of glass or quartz vessels at an intensity exceeding that of the irradiated solutions. Moreover, the time interval between the end of irradiation and the start of the measurement of chemiluminescence must be short and the photometric sensitivity of the apparatus high. Uncooled photomultipliers (PM) should be used. The best results have been obtained with an experimental setup in which the flow of liquid is first irradiated in a cuvette and then enters a vessel placed near the photocathode of a head-on PM. Thin polyethylene tubes 1.5—2 mm in diameter are used in order to increase the linear rate of the flow. In this manner chemiluminescence processes can be detected within a few seconds after irradiation. The experimental results obtained so far (UV- and X-ray induced afterglow of solutions of proteins, polymers, amino acids and dyes) indicate that the chemiluminescence processes are very prolonged; accordingly, the time interval between irradiation and measurement can be 10—15 seconds. The use of a jet injector with a high flow rate leads to a high expenditure of solution, which is not always desirable. It was found that with a low irradiation rate the experiment can be carried out within a closed system which allows recycling of the flow. The minimum volume of liquid necessary for the operation of a closed system, (i. e., the combined volume of the irradiation and measurement cuvettes, the pump and the polyethylene tubing) is 15—20 ml.

The liquid is propelled by piston-type micropumps. The flow rate can be controlled by changing the frequency of the audio-frequency oscillator feeding the pump motor. Centrifugal pumps provide a high volume rate, but their application is limited by the close relationship between the rate and the resistance of the tubing.

A reciprocating movement of the liquid was used in another variation of the test. In this case the outlet of the measuring vessel was equipped with a syringe operated by an electric motor. The liquid, 4—5 ml in volume, was pumped over from one vessel to the other within 3—5 seconds. The vessels for irradiation and measurement were thermostatically controlled and varied greatly in size and form. The optimal dimensions of the measuring vessel depend on the size of the multiplier photocathode. Vessels with a diameter of 35 mm and a thickness of 3—5 mm were used for measurement. The thickness of the thermostatic coat is not important. The irradiation was carried out in two types of vessels, open and closed at the top.   The housing of the measuring cuvette represented a continuation of the PM housing.   The housing inlets must not let in light.   This requirement was met by constructing the transmission system so that the polyethylene tubing had 3–4 bends in it, or by threading the tube into a thin vacuum tube 15–16 cm long.   For X-ray work the PM housing was screened with 5–6 mm of lead and was placed outside the beam of the X-ray tube.

The electrical unit of the apparatus was conventional.   The PM was fed by stabilized rectifiers such as VS-9, VS-16 or SVV-10.   The photocurrent of the multiplier was fed to a direct-current amplifier.   The kinetic curves were recorded by means of an EPP-09 electronic potentiometer. Some of the experiments were done using the EPPV-60 high-resistance potentiometer.   The photocurrent of the multiplier is of the order of $10^{-10} - 10^{-12}$ ampere.

The determination of the experimental quantum and ionization yields of chemiluminescence and other associated parameters requires absolute measurements of intensity of chemiluminescence as well as knowledge of the irradiation dose rate (the irradiation intensity in the case of light).

The absolute quantum sensitivity of the photometric apparatus was determined as described in [75, 76] using a liquid radioactive phosphor with a known emission spectrum and light yield.   This approach obviates the necessity to take the geometry of the photometric system into account. For the calibration it is also necessary to know the spectral sensitivity of the PM and the chemiluminescence spectrum.

The apparatus is equipped with a "standard signal" in the form of an ampule with a radioactive pigment and a seal.   All measurements refer to the magnitude I_{ss} of this signal.   In our experimental systems deviation of the recording pen following the introduction of the standard signal is equivalent to a light source with a spectrum similar to the chemiluminescence spectrum of proteins and an intensity of about 10^4 quanta of chemiluminescence/(ml.sec) [53].

The chemical dosimetry of the X- and gamma-ray sources is carried out by the conventional method (ferrous sulfate dosimetry).   The actinic power of the UV sources was measured by means of a ferrous oxalate dosimeter as described in [77, 78], and of the visible light according to the procedure outlined in [79].   The spectral analysis of induced chemiluminescence is very difficult because of its low intensity.   There are two common procedures for analyzing the spectra, one involving the use of high-power monochromators and the other light filters [80].   The first procedure yields a continuous spectrum, the second – a discontinuous one.   Both methods have been used in the studies discussed below;   reliable information on the emitter is unfortunately limited because no existing apparatus for spectral analysis provides data on the relative intensity of luminescence with an error of less than 15%.   In conclusion, the equipment for the analysis of induced chemiluminescence is very simple (for integral measurements). It consists essentially of standard components;   the difficulties lie in the selection of the PM and in the spectral analysis.

Apart from the chemiluminescence observed in solutions exposed to X-rays or gamma rays, UV or visible light in the presence of sensitizers, there exists an activated chemiluminescence in which the intensity of glow rises sharply with the introduction of certain dyes serving as activators. This effect is purely physical and is associated with the last stage of the process; the interaction between the dye and the molecule of the excited-electron product involves a migration of energy, and the de-excitation takes place from the electron levels of the activator.

All these categories of phenomena are closely interrelated and often differ from one another solely by the primary processes. The kinetic curves of the afterglow are in some cases exponential, in others follow a more complex course. The time τ during which the glow decays by a factor of e varies in different compounds from a few seconds to several tens of minutes. The initial intensity of the chemiluminescence depends on the experimental conditions and amounts to 10^3-10^5 quanta of chemiluminescence/(ml.sec). The chemiluminescence spectra of protein solutions form wide bands from 380 to 520 nm with a peak at about 440–460 nm.

The X-ray induced chemiluminescence of certain compounds has been described in the literature [14, 17]. It was found that a large variety of compounds emit an afterglow after exposure to unfiltered X-rays for 5 minutes (dose rate 450 rad/min). This applies to various inorganic compounds such as sodium chloride, potassium chloride and potassium bromide at concentrations higher than 1%. Potassium iodide and iron sulfate show a negligible chemiluminescence. A slight afterglow was observed in solutions of sodium hydroxide and potassium hydroxide, and in mono- and dibasic sodium phosphate. The luminescence of solutions or inorganic compounds has a low intensity and decays within 1–3 minutes. An afterglow was detected in dulcitol, arabitol, saccharose (low intensity), glucose, heparin, DNA and RNA (decay time 5–10 minutes). Chemiluminescence is practically absent in distilled water.

More thorough studies were made of the X-ray induced chemiluminescence of 19 amino acids, 8 peptides and 12 proteins, in which the chemiluminescence intensity is high in comparison with other organic compounds. The following parameters were selected to characterize X-ray induced chemiluminescence: the relative initial intensity $\frac{I_0}{I_{ss}}$, representing the initial rate of the process responsible for luminescence; the time τ during which the intensity decays by a factor of e; and the light sum (the area below the afterglow curve), which is proportional to the ionization yield and to the total yield of chemiluminescence. Table 3.2 shows experimental data on the chemiluminescence of amino acids at a concentration of $3 \cdot 10^{-2}$ M. The light sum of tryptophan is taken as 100.

It is evident from Table 3.2 that the X-ray induced chemiluminescence of tryptophan yields the highest values of light sum and intensity. These values are larger by 1–1.5 orders of magnitude than those for the other amino acids; τ varies much less.

Of the peptide solutions examined, glycyltryptophan showed the highest intensity of afterglow. The afterglow of solutions of alanyl-glycine, glycyl-leucine-valine is lower by one order, while the X-ray induced chemiluminescence of gly-gly, gly-leu, gly-tyr, gly-phen and glutathione is very weak.

Kinetic curves of X-ray induced chemiluminescence were obtained for the solutions of 12 proteins at a concentration of $5.8 \cdot 10^{-5}$ M. The curves follow a more or less uniform course. The decay of glow is exponential in some proteins but more complex in others. Table 3.3 shows data on the X-ray induced chemiluminescence of solutions of various proteins. The light sum of human serum albumin (HSA) is taken as unity.

TABLE 3.2. X-ray induced chemiluminescence of amino acid solutions

Amino acid	Light sum	τ, min	$\dfrac{I_o}{I_{ss}}$	Amino acid	Light sum	τ, min	$\dfrac{I_o}{I_{ss}}$
Tryptophan	100	5.5	4.83	Tyrosine	2.2	3.2	0.02
Arginine	7.3	2.0	0.46	Glycine	1.9	2.2	0.31
Isoleucine	5.4	2.0	0.45	Proline	1.3	1.5	0.71
Lysine	4.8	3.2	0.34	Valine	1.2	0.6	0.26
Alanine	4.2	1.3	0.40	Histidine	0.3	1.3	0.60
Aspartic acid	3.9	1.0	0.50	Glutamic acid	0.0	0.8	0.40
Serine	3.7	3.5	0.22	Methionine	0.0	–	0.00
Leucine	2.8	1.4	0.40	Cystine	0.0	–	0.00
Threonine	2.6	0.9	0.53	Cysteine	0.0	–	0.00
Phenylalanine	2.1	1.7	0.22				

The plant proteins papain and glycinin were dissolved at a concentration of 0.2% in a 10% solution of sodium chloride. Hemoglobin and cytochrome C show a very low intensity afterglow, possibly because of a reabsorption of the light of chemiluminescence.

TABLE 3.3. X-ray induced chemiluminescence of protein solutions

Protein	Light sum	τ, min	$\dfrac{I_o}{I_{ss}}$
HSA	1	2.3	1.67
DNase	0.93	4.1	1.16
Hyaluronidase	0.88	3.8	1.13
Papain	0.86	3.0	1.29
Lactate dehydrogenase (LDH)	0.61	2.1	1.08
Trypsin	0.52	2.5	0.72
Bovine serum albumin (BSA)	0.52	1.4	1.15
Egg albumin	0.45	2.6	0.72
Glycinin	0.37	1.2	0.94
RNase	0.27	2.5	0.50
Cytochrome C	0.03	2.4	0.05
Hemoglobin	0.02	1.5	0.06

The light sum is highest in HSA and lowest in the RNase solution. A comparison was made between the X-ray induced chemiluminescence and amino acid content of the protein solutions. No relationship exists between the light sum and the molecular weight of the protein. The total yield of the glow is lower in proteins which do not contain tryptophan (glycinin and RNase). A definite relationship is evident between the chemiluminescence and the proportion of tryptophan in the molecule but not the total tryptophan content. This indicates a strong competition between tryptophan residues and other groups in the protein for the radicals formed during radiolysis.

Assuming that accessibility of the tryptophan residues to the active products of radiolysis of water varies from one protein to another, it seems that these residues are superficial in DNase and HSA but concealed in other proteins.

All other conditions being equal, the magnitude of τ depends on the prospects for the "withdrawal" of the free radicals by neighboring acceptor groups of the protein, possibly sulfur-containing amino acid residues situated near the site of the free radicals. It was found that τ tends to decrease with a rise in the number of disulfide bonds. Thus, τ equals 4.1 minutes in DNase (no disulfide bonds) and 1.4 minutes in BSA (16 disulfide bonds); the chemiluminescence of these proteins differs only slightly in initial intensity.

The ionization yield of the chemiluminescence of glycyl-tryptophan is $1.4 \cdot 10^{-8}$ quanta of chemiluminescence/ionization [16], which is more than the corresponding values for tryptophan and HSA.

The X-ray induced chemiluminescence of HSA solution was investigated by means of models [81]. Two models were examined. The first consisted of a solution of amino acids equivalent to HSA in composition; as might be expected, the luminescence obtained was weaker than in the protein because of the competition between the different amino acids for active products of the radiolysis of water. The second model was used to study the afterglow of the individual amino acids at concentrations corresponding to their concentrations in the protein. The latter approach ruled out competition between amino acids for radicals resulting from radiolysis of water as well as reactions between radicals formed from the different amino acids. The X-ray induced chemiluminescence of the protein was compared with the sum of the afterglow produced by the amino acids. It was found that the second model characterizes the luminescence of the protein more accurately than the first.

A study was made of the eosin-sensitized photochemiluminescence of solutions of compounds such as amino acids, peptides, sugars, nucleic acids, and synthetic polymers [68]. The solutions were irradiated for 30 seconds with a DRSh-250 mercury lamp through light filters isolating the 546 nm line; the actinic power of the source was measured and the absorbed dose calculated (eosin concentration $1.82 \cdot 10^{-5}$ M).

An eosin-sensitized photochemiluminescence was detected in the solutions of 18 amino acids at a concentration of $2.72 \cdot 10^{-2}$ M. Tryptophan has the greatest quantum yield ψ of chemiluminescence — $5.6 \cdot 10^{10}$ quanta of chemiluminescence/absorbed quantum; the corresponding values for phenylalanine, arginine and threonine are lower by an order of magnitude (respectively, $6 \cdot 10^{-11}$, $5 \cdot 10^{-11}$, and $2.4 \cdot 10^{-11}$). The yield of sensitized chemiluminescence of isoleucine, serine, lysine, glycine, alanine, methionine, leucine, histidine, proline, cysteine, tyrosine, glutamic acid and aspartic acid is less than $2 \cdot 10^{-11}$ (the lowest value which could be measured with the apparatus used).

The quantum yields of the sensitized chemiluminescence of the peptides (concentration $4.6 \cdot 10^{-3}$ M) were found to be as follows: glycyl-tryptophan — $1.3 \cdot 10^{-8}$; glycyl-phenylalanine — $1 \cdot 10^{-10}$; glycyl-tyrosine — $4 \cdot 10^{-11}$; alanine-glycine — $2 \cdot 10^{-11}$; glycyl-leucine-valine — $6 \cdot 10^{-11}$. The yields of other peptides (ala-ala, trigly-gly, gly-asp, gly-val, ala-met) are less than $2 \cdot 10^{-11}$. Thus, the yield is highest in glycyl-tryptophan and lower by 2–3 orders of magnitude in peptides which do not contain tryptophan.

The irradiation of protein solutions resulted in strong photochemiluminescence which persisted for many minutes after exposure to light.

TABLE 3.4. Sensitized photochemiluminescence of protein solutions (protein concentration $5.27 \cdot 10^{-5}$ M)

Protein	$\psi, 10^9$	x	$\frac{\psi}{x}, 10^9$
BSA	735	302	2.4
HSA	664	89	7.4
Fibrinogen	150	15.4	9.8
Globulin	9	11.4	7.8
DNase	13	–	–
Egg albumin	10	24.4	0.4
Trypsin	5.8	1	5.8
Lysozyme	5.2	13.2	0.4
LDH	3.2	11.4	0.3
Hyaluronidase	2.8	4	0.7
Chymotrypsin	2	1.6	1.2
RNase	1.0	1.97	0.5
Trypsin inhibitor	0.25	–	–

It is evident from Table 3.4 that the quantum yield of sensitized photochemiluminescence in most proteins is higher than in amino acids and
peptides.   It has been shown that eosin can serve as activator, increasing
the photochemiluminescence of irradiated proteins [45].   The eosin present
in a system thus functions not only as a sensitizer but also as an activator.
Experiments were performed to determine the amplification factor x according to intensification of photochemiluminescence of protein solutions
when exposed to UV light after irradiation.   The values obtained for the
amplification factor are also shown in Table 3.4 together with the ratio
between the quantum yield and the amplification factor, which represents
the true yield of the chemiluminescence.   The values of this corrected
yield in different proteins vary less than the experimental values obtained.
However, the determination of the amplification factor according to the UV-
chemiluminescence of proteins is hampered by all kinds of difficulties and
the values obtained for the true yield are not very reliable.
    Sensitized photochemiluminescence has been observed in nucleic acids,
sugars and polymers.
    The yield in these compounds is lower than that obtained from proteins.
It is evident from the data obtained that sensitized chemiluminescence is a
widespread phenomenon.   Proteins show the highest quantum yield of
chemiluminescence; of the peptides and amino acids examined, the greatest
values were obtained with glycyl-tryptophan and tryptophan, respectively.
    Chemiluminescence also appears as a result of the direct exposure of
various substances to UV light.   The quantum yield of photochemiluminescence has so far been determined only for a small number of compounds,
probably because of technical difficulties; for glycyl-tryptophan it is
$2 \cdot 10^{-9}$ quanta of chemiluminescence/UV quantum [36], and for human serum

albumin $3 \cdot 10^{-11}$ quanta [53]. The yield of induced chemiluminescence in other compounds is lower than in proteins. No quantitative study of the effects of monochromatic UV irradiation on a large range of proteins with reference to the absorption by tryptophyl residues, tyrosine, phenylalanine and cystine has yet been undertaken.

TABLE 3.5. The sensitized photochemiluminescence of different compounds

Compound	Concentration, $\%$	$\psi,\ 10^{10}$	x	$\dfrac{\psi}{x},\ 10^{10}$
DNA	0.2	17	1	17
RNA	0.2	1.0	1	1.0
Glucose	2	0.4	1	0.4
Saccharose	2	0.2	1	0.2
Heparin	1	2.7	1	2.7
Polyethylene oxide	0.5	3.0	1.45	2.0
Polyvinylpyrrolidone	1	20	3.4	6.0

Of particular interest are the data relating to the excitation spectra of chemiluminescence. Here it is difficult to obtain adequately resolved spectra because of the low yield of the luminescence.

Using light filters with a short wave transmission cutoff, it was shown that wavelengths of less than 300 nm are involved in the UV excitation [27, 40]. We obtained excitation spectra of the photochemiluminescence of a solution of glycyl-tryptophan and of the sensitized chemiluminescence of HSA solution. The sources of excitation energy in these experiments were a DRSh-250 mercury lamp and a DKSSh-200 xenon lamp with the beam passed through a 3MR-3 quartz monochromator; a wide slit (spectral width about 15 nm) was used in the experiments, with the intensity of light excitation kept constant over the whole range of wavelengths; excitation intensity was measured by means of a photocell with luminescence conversion.

A qualitative correlation was found between the excitation and absorption spectra. Further research is necessary to obtain highly resolved excitation spectra.

MECHANISM OF THE RADIATION-INDUCED CHEMILUMINESCENCE OF SOLUTIONS OF PROTEINS AND PEPTIDES

Initial research on the prolonged luminescence of protein solutions exposed to UV-irradiation gave rise to the following hypotheses as to the causes of this phenomenon.

1. Luminescence is associated with very persistent excited metastable states lasting more than 100 seconds. This phenomenon cannot be due to phosphorescence of the protein since the lifetime of the triplet states of a protein in solution does not exceed $10^{-3}-10^{-4}$ seconds [26, 38].

2. Luminescence results from a recombination or disproportionation of the free radicals arising from the protein on irradiation of the solution [41].

3. Luminescence is associated with reactions of the peroxide compounds formed during the irradiation [40].

Experiments were carried out to determine the effect of oxygen on the appearance of the prolonged afterglow [41]. It was found that luminescence does not appear in the absence of oxygen and thus the first hypothesis can be ruled out. The second and third assumptions necessitated setting up special experiments to study the kinetics of the afterglow. This work was done with HSA solutions. The kinetic curves of the afterglow revealed a first-order rate constant for the decay of luminescence. It was found that the magnitude of the rate constant (i.e., the form of the kinetic curve of the afterglow) is independent of the protein concentration within the range of 0.03—1.0%. Two series of experiments were performed; in one series, protein solutions of different concentration were irradiated, while in the other series the solutions were diluted after irradiation. The rate constant for the decay of luminescence had an activation energy of 10 ± 2 cal/mole — an indication that the afterglow cannot be due to the thermal decomposition of peroxide compounds. A chemiluminescence is also observed during the reaction of protein with peroxides [82], but its kinetic characteristics differ strongly from those of photochemiluminescence. It follows from these and other considerations that the prolonged afterglow can hardly be due to a peroxide mechanism.

Various compounds were tested for their effect on the prolonged afterglow. Both inhibitors of free-radical processes, i.e., compounds reacting readily with peroxide radicals but only slightly with peroxides, and compounds reacting with peroxides and radicals alike, were used in these experiments. It was found that all these compounds affect the rate constant of the afterglow and that the relationship between the rate constant and the concentration of the additive is linear. Mathematical analysis of the kinetic schemes shows that such a relationship is possible if the process results from disproportionation of the free peroxide radicals generated by irradiation [41]. Indirect evidence for this conclusion was supplied by a parallel study of chemiluminescence, electron paramagnetic resonance (EPR) and oxygen absorption of dry proteins in a solid phase; these experiments showed that chemiluminescence results from an oxidative recombination of the radicals of the irradiated protein. Our numerous attempts to detect an EPR signal in the solution following UV radiation of the protein proved unsuccessful; as will be shown below, the possible stationary concentration of radicals is far below the sensitivity of the EPR.

The experimental finding that the rate constant for decay of chemiluminescence does not depend on the concentration is very strange since the disproportionation of the free peroxide radicals must by its very nature, be a second-order reaction. It was apparent that chemiluminescence is a complex process comprising several reactions. A kinetic analysis was carried out in order to determine the sequence of radical reactions responsible for the afterglow [47].

This study began with an analysis of the relationship between the intensity of glow and the concentration of active photolytic products (the solution was diluted after irradiation). For BSA solution the relationship is linear. The order of the reaction with respect to concentration was found to be 0.99 + 0.17. The kinetics of the buildup of luminescence at different radiation dose rates was then studied in order to obtain additional data on the photochemiluminescence process. It was found that the intensity of the luminescence grows exponentially with time to saturation. The magnitude of this limiting value of the intensity of chemiluminescence increases with the intensity of UV irradiation. The magnitude of the rate constant for the buildup of chemiluminescence increases linearly with the radiation intensity.

The effect of the UV radiation intensity on other parameters of the curves characterizing the buildup of chemiluminescence was also determined. Several kinetic schemes were examined in order to determine the mechanism involved, and the calculated data compared with the experimental findings. It became clear that only one of the simplest schemes can explain all the available experimental evidence. According to this scheme the peroxide radicals generated by irradiation isomerize into secondary free radicals and the subsequent disproportionation of these causes chemiluminescence. At the same time, a new process was detected — a radiation-induced destruction of radicals.

It is evident from the available data that the limiting stage of the process is a first order reaction with respect to the primary radicals of the protein. The nature of the isomerization process remains obscure, and there is no obvious explanation as to why the direct disproportionation of the peroxide radicals is not detected. It is noteworthy that illumination not only generates free radicals but also destroys them. The radicals are destroyed by light with a wavelength of less than 300 nm.

A number of hypotheses may be advanced to explain this phenomenon. Possibly the protein molecule produces a second radical which undergoes instantaneous recombination [83]; alternatively, the incident light may cause direct decomposition of the radical; a third possibility is that destruction of the radicals requires absorption of light by chromophores of the protein.

A general scheme for chemiluminescence of proteins may be drawn up on the basis of the available data [41, 47]. However, the nature of the protein groups responsible for the particular stages in the complex process of chemiluminescence remains unknown. An aromatic amino acid — tryptophan — was regarded as a crucial participant in one or several stages of photochemiluminescence. A comparative study of fluorescence and photochemiluminescence of BSA solutions was carried out in order to determine the role of the tryptophyl residues [46]. The tryptophan content of proteins can be determined by fluorescence at 320—360 nm on excitation at a wavelength of less than 300 nm. A flow fluorescence detector was set up and simultaneous measurements were made of the intensity of chemiluminescence and the breakdown of tryptophan. As a result of UV irradiation [about $5 \cdot 10^{15}$ quanta/(ml· sec)] the tryptophan content of the protein decreases and the kinetic curve showing the combustion of tryptophan follows a first order equation [84] with a rate constant of $3.4 \cdot 10^{-3} \sec^{-1}$. The stationary intensity of the chemiluminescence at first rises, reaching a peak within 1 minute and then steadily declines. The descending part of the curve similarly shows an exponential relationship with a rate constant of $3.3 \cdot 10^{-3}$ sec^{-1}, which corresponds to the rate constant of the photodecomposition of tryptophan.

The manner in which the addition of cysteine (a potent acceptor of free radicals from protein) and eosin (an activator) affects the breakdown of tryptophan and the chemiluminescence was also established. These additives were introduced at a concentration sufficiently low for their effect on the excitation process to be neglected. It was found that neither cystine nor eosin alter the rate of tryptophan photolysis; the experimentally determined rate constants all coincided at a value of $3.3 \cdot 10^{-3}$ sec^{-1}. At the same time these additives strongly influence the intensity of chemiluminescence (eosin causes a 15-fold increase, cysteine a 1.5-fold decrease), while the rate constant for the decay of luminescence remains unchanged at $3.3 \cdot 10^{-3}$ sec^{-1}.

The processes involving tryptophan may be outlined as follows. Light is absorbed by tryptophan residues of the protein, which as a result decompose to yield free radicals. Since the rate constants for decomposition and chemiluminescence are equal, this process can be regarded as the only pathway of tryptophan breakdown under the experimental conditions setup (the expenditure of tryptophan would be faster if there were an additional pathway). One minute after the beginning of irradiation the system enters a quasistationary state in which the radicals appear and vanish at the same rate.

The decay of chemiluminescence during irradiation can be attributed to the decreasing rate of radical formation, which results in turn from the depletion of tryptophan. The form of the kinetic curves of the afterglow remains unchanged, however. This may mean that there is no migration of energy from the excited product to tryptophan. If there were such a transfer of energy from the excited product to tryptophan, the decrease in its concentration would cause a faster decay of the chemiluminescence owing to the reduced rate of radical formation and the "burning-up" of activator. This signifies that the emission centers are not the tryptophan residues but some other products of the disproportionation of radicals. On this basis it may be said that tryptophan residues in the protein perform a crucial function in the excitation of chemiluminescence as the source of free radicals, generated by the photolysis of the amino acid. The subsequent reactions do not affect the tryptophyl residues, and tryptophan residues are not involved in the emission of light by chemiluminescence.

The discovery of the gross mechanism of the photoluminescence of protein solutions cleared the way for a quantitative study of the processes taking place from the moment of absorption of the radiation energy to the moment at which chemiluminescence is recorded, enabling identification and assessment of the main parameters of photochemiluminescence of protein solutions [53]. Before considering the results of such studies we must note the following designations: ψ — the experimental quantum yield of chemiluminescence, defined as the number of quanta of chemiluminescence per quantum of energy absorbed; φ — the quantum yield of radicals, i. e., the number of radicals formed per quantum absorbed; η — the chemiluminescence yield, expressed as the number of chemiluminescence quanta divided by the number of events leading to chemiluminescence; η_{exc} — the number of excited states per reaction event; η_P — the emission yield of the excited product, expressed as the number of chemiluminescence quanta per excited state.

The kinetics of the processes involved in photochemiluminescence of protein solutions can be outlined as follows on the basis of existing experimental data [41, 46, 47].

1. Formation of free radicals: $P \xrightarrow{h\nu, O_2} R\cdot$. It has been established that peroxide radicals $R\cdot$ are formed by decomposition of tryptophan. The rate of formation of radicals depends on their quantum yield and the intensity I_0 of absorbed light; it is proportional to the product of the breakdown rate of tryptophan multiplied by its concentration.

2. Reactions of the radicals. The main process here is destruction of the radicals by the radiation:

$$R\cdot \xrightarrow{I_0} \text{products}.$$

The rate of this process is expressed by the product $k_1' \cdot I_0 \; [R\cdot]$, where $k_1 = k_1' \cdot I_0$ is the effective, experimentally determined rate constant. The other reactions are isomerization of the radicals $R\cdot \xrightarrow{k_0} r$ and disproportionation of the secondary radicals $2r \cdot \rightarrow P$. Molecules of the product P in an excited-electron state are formed at a low yield η_{exc}.

3. The process $P^* \rightarrow P + h\nu_{ch}$, characterized by the yield η_P.

On the basis of the above scheme the corresponding kinetic equations have been derived linking all the major parameters of chemiluminescence. The parameters were determined by kinetic studies of the buildup and decay of chemiluminescence (determinations of k_0 and k_1) and measurements of the absolute intensity of the glow and the intensity of the incident light. The following quantitative parameters were obtained [45]:

$$\psi = 2.6 \cdot 10^{-11} \text{ quanta/UV quantum;}$$

$$\varphi = 4.7 \cdot 10^{-3} \text{ radical/UV quantum;}$$

$$\eta = 1.1 \cdot 10^{-7} \text{ quanta of chemiluminescence/radical;}$$

$$\eta_{exc} = 1.2 \cdot 10^{-4} \text{ excited states/radical;}$$

$$\eta_P = 0.9 \cdot 10^{-3} \text{ quanta of chemiluminescence/excited state.}$$

The kinetic rate constants have the following values: $k_0 = 2.3 \cdot 10^{-3} \text{ sec}^{-1}$; $k_1' = 0.6 \cdot 10^{-17} \text{ ml/UV quantum}$. On the basis of these data it is possible to calculate a number of important values, notably the initial stationary concentration of peroxide radicals; the latter value was found to be $6.4 \cdot 10^{14}$ radicals/ml, and the maximum possible concentration (owing to destruction of the radicals by radiation) $8 \cdot 10^{14}$. It is evident from these data that concentration of radicals can hardly attain a level measurable by methods such as EPR.

The values obtained for the yield and constants are reasonable and in some cases comparable with those previously established — an indication that the proposed mechanism is correct. Each of the parameters determined in [53] characterizes highly complex effects and processes which are just beginning to be studied. For example, the value φ is associated with processes involved in the excitation of tryptophan and the formation of certain intermediate forms reacting with oxygen to yield peroxide radicals; the value k_0 is linked with a still obscure reaction comprising the slow isomerization of radicals, etc.

Thus, the mechanism of photochemiluminescence of protein solutions has been successfully established and expressed in kinetic and photo-chemical terms.

Of particular interest is the physical nature of the excited states of the product generated in the process. Research in this field has two aims — to establish the chemical nature of the emitter and the characteristics of the chemiluminescence. Many workers are interested in the prospects for chemically generating such excited states, which are difficult to produce in a solution on photoexcitation. Two approaches are commonly used, namely the introduction of either quenching agents or activators. In the photo-chemiluminescence of proteins (just as with any other chemiluminescence) it is difficult to select substances which influence only one particular stage of the process without altering the chemical process as a whole.

The effect of more than 20 dyes on the photochemiluminescence of HSA solutions was examined [45]. It was found that only fluorescein and its derivatives (eosin, erythrosine) increase the intensity of glow. The addition of eosin increases the photochemiluminescence of many proteins (p. 84). It may be assumed that eosin participates directly in the chemical reactions leading to luminescence, reveals new luminescence centers, or functions as an activator, which does not influence the chemical process but acts through a transfer of energy from the luminescence center to itself. Detailed kinetic studies of the activated luminescence of protein solutions (time course, temperature dependence, effects of pH cysteine as an additive, etc.) have shown that eosin does not influence the chemical process. The spectra of activated chemiluminescence and photoluminescence of the protein-dye systems were found to coincide. It was found that introduction of the dye lowers the intensity of chemiluminescence of the protein and increases that of the protein-dye system [50]. These experiments indicate that there is a migration of energy from the molecule of the excited product to eosin. By studying the relationship between the amplification factor and the con-centration of the stain it has been possible to calculate the quantum yield of emission from the excited product $(0.9 \cdot 10^{-3})$.

Initially, it was assumed that the mechanism of amplification consisted in an energy transfer during collisions [85] — a view apparently confirmed by the concentration effect. More thorough studies have established, how-ever, that the energy transfer takes place in the protein-dye complex, and that the amplification factor depends on parameters such as the protein-dye bonding constant, concentration of the protein, the luminescence yield of the dye, the rate constant of energy transfer, etc. [86]. To our knowledge, the effect of quenching agents on photochemiluminescence of proteins has not been studied systematically [87].

We may now turn to the nature of the chemiluminescence center in the protein molecule. Direct kinetic data on this topic are not yet available. It was shown on a glycyl-tryptophan solution, however, that the center appears to be kynurenine. The chemiluminescence spectrum of proteins closely resembles the photoluminescence spectrum obtained upon excitation, by means of light with a wavelength of 365 nm, of a previously irradiated protein in which kynurenine-type products have accumulated to a sufficient extent. This finding is supported by indirect evidence (formation of kynurenine during sensitized oxidation of a number of proteins). Thus, there are sufficient grounds for regarding kynurenine as the center of induced chemiluminescence of protein solutions. This assumption may soon be confirmed by direct experimental data.

Research on photochemiluminescence of proteins faces great difficulties in the interpretation of both quantitative data and conjectures owing to the complexity of the protein molecule, uncertainty as to the position of trypto-phan residues in the protein molecule and the effect of neighboring groups on these.

For this reason photochemiluminescence of proteins had to be studied in simpler models. Glycyl-tryptophan was one of the models chosen. The analysis of a simpler system is also of considerable interest in itself because it gives a deeper insight into the mechanism of induced chemi-luminescence. Below we shall briefly review experimental results obtained with glycyl-tryptophan solutions on UV irradiation, sensitized exposure to light, and gamma irradiation. Many of the kinetic characteristics of chemiluminescence induced in the solutions were found to be identical for all these diverse types of radiation. The data on the kinetics and yield of chemiluminescence will be discussed first [36].

All the kinetic curves of the afterglow of the peptide solutions are ex-ponential. The first order rate constants of the decay of luminescence are $(1.5-2)\cdot 10^{-2}$ sec^{-1}. This appeared strange (as in the case of protein) since one might expect the process to follow a strictly bimolecular course. The exponential form of the kinetic curves could be attributed to a strong inter-action between the free radicals and the peptide itself, which would mean that two reactions take place in parallel — disproportionation of radicals and their withdrawal by a reaction with the original compound. In this case the rate constant must increase with concentration of the peptide.

However, the rate constant does not depend on concentration of glycyl-tryptophan within a 500-fold range of the latter. The addition of 50 times the original concentration of the peptide to the irradiated solution does not affect the course of the kinetic curve of the afterglow. These findings indicate that the rate of the afterglow process is limited by a stage charac-terized by first order kinetics with respect to concentration of the radicals formed, just as in the case of proteins. On this basis the following scheme of processes was considered. UV irradiation with a quantum yield φ gener-ates free peroxide radicals $R\cdot$. These radicals can isomerize into second-ary radicals, $R\cdot \xrightarrow{k_0} r\cdot$, which disproportionate to yield product molecules in an excited-electron state; the light of chemiluminescence is emitted from levels of these molecules. In addition, there is a radiation-induced destruction of the radicals. For a relatively brief time of irradiation τ_{irr} this scheme of events leads to the following equation for chemiluminescence intensity:

$$J = \eta\varphi I_{irr}\, k_0\tau_{irr}.$$

From this equation the relationship between the intensity of chemi-luminescence and that of irradiation is obviously linear. Experiments have indeed shown that chemiluminescence intensity is directly proportional to irradiation intensity within a more than 30-fold range of the latter. The value of k_0 meanwhile remains constant. By calculating the tangent of the angle of inclination of the straight line it is possible to determine the quantum yield of chemiluminescence:

$$\psi = \eta\varphi = 1.9 \cdot 10^{-9} \text{quanta of chemiluminescence/UV quantum.}$$

The rate of breakdown of glycyl-tryptophan and the quantum yield of photolysis of the peptide, φ, were also determined; $\varphi = 3.2 \cdot 10^{-2}$ mole-cules/UV quantum.

It was assumed [36] that the breakdown necessarily involves free radicals and that one radical is produced per molecule. From the values of ψ and φ it is possible to calculate the chemiluminescence yield:

$$\eta = 6.6 \cdot 10^{-7} \text{ quanta/molecule.}$$

We shall not dwell here on the various complications [36] arising from the influence of photolytic products on the intensity of chemiluminescence. The rate constant of radiation-induced destruction of radicals, determined from the kinetic curve of chemiluminescence buildup, was found to be $0.5 \cdot 10^{-17}$ ml/UV quantum.

Comparison of these results with the corresponding data for HSA reveals marked differences in the efficiency of generation of radicals and in the rate constants for isomerization. The former difference can be attributed to the fact that in proteins both tryptophan and tyrosine residues absorb UV radiation. As will be shown below, the discrepancy in rate constants for decay of the glow probably results from differences in the solvation processes.

The values of the main parameters of photochemiluminescence of glycyl-tryptophan solutions depend on the experimental conditions. A study was made of the effect of pH on the quantum yield of photolysis of the peptide, intensity of chemiluminescence, magnitude of the rate constant of izomeriza-tion of radicals, and chemiluminescence yield [35]. It was found that pH strongly influences intensity of chemiluminescence: there is practically no glow at pH values up to 5.5, then the glow intensity rises sharply in the region of pH 6—8.5 to a peak at pH 9—10.5 and declines again in the more alkaline region. The other parameters are practically constant in the pH 5—13 range.

The values of chemiluminescence yield were also determined since changes in this parameter are responsible for the differences in intensity. The effect of pH on the yield can be represented by two titration curves with pK values of 7.4 and 11.5. The changes in chemiluminescence yield are attributed mainly to protonation and deprotonation of the excited state of the product.

Radiobiologists are naturally more concerned with the indirect effects of radiation, i. e., the processes of formation of organic free radicals by reactions involving the active products of water radiolysis and the active products of other compounds in the living cell. These topics are discussed in [16, 69], which deal respectively with gamma chemiluminescence and sensitized chemiluminescence of glycyl-tryptophan solutions. Both studies indicate that peptide yields radicals by reactions with active intermediate particles; their main contribution is to show the possibility of obtaining data on such rapid reactions.

Study [16] of the kinetic characteristics of gamma chemiluminescence of the same peptide solutions has shown that the kinetics of the glow decay and the chemiluminescence yield coincide with the values obtained in the case of UV excitation. The ionization yield was found to be $1.4 \cdot 10^{-8}$ quanta/ionization, the radiochemical yield of breakdown of the peptide 3.3 molecules/100 eV, which corresponds with the findings reported in [88], and the chemiluminescence yield is of the order of 10^{-8}. The intensity of chemiluminescence rises with concentration of the peptide from 10^{-7} to $3 \cdot 10^{-3}$ M — rapidly at first, then very slowly at a concentration of $3 \cdot 10^{-5}$ M, then again more quickly with a further rise in the concentration apparently as a result of the involvement of "excited" water molecules in the process. The course of the curve up to 10^{-4} M suggested that the afterglow could be attributed to an indirect effect of radiation. Experiments involving addition of small amounts of methanol — a good acceptor of hydroxyl radicals — were performed in order to identify the radiolytic products of water responsible for the observed effects. It was found that methanol sharply lowers the intensity of chemiluminescence. Quantitative analysis of the effect of concentration has enabled determination of the ratio between the rate constants of reactions of hydroxyl radicals with methanol and with glycyl-tryptophan (0.054). The elementary rate constant of the reaction between the hydroxyl radical and glycyl-tryptophan was accordingly found to be $1.2 \cdot 10^{10}$ M^{-1} sec, which agrees with the findings reported in [89].

It is clear, therefore, that the only difference between gamma chemiluminescence and photochemiluminescence of glycyl-tryptophan solutions lies in the primary mechanism of generation of free radicals from the peptide. The emission of light results from reactions between hydroxyl radicals and glycyl-tryptophan molecules. The entire process of gamma chemiluminescence of solutions of this peptide can be characterized in quantitative terms on the basis of the values obtained.

Sensitized chemiluminescence of glycyl-tryptophan solutions is discussed in [69]. A solution of this peptide containing eosin was illuminated in the absorption range of eosin. The intensity of chemiluminescence increases with the concentrations of eosin and the peptide. The relationship between this intensity and concentration of the peptide is typical of an indirect effect. The authors explained these processes by a scheme covering not only the usual stages (see p. 92) but also formation and decomposition of the active products of photolysis of eosin. Quantitative analysis of experimental data on the basis of this scheme enabled determination of the ratio between the rate constant of deactivation of the active products and the rate constant of their reaction with glycyl-tryptophan ($2.8 \cdot 10^{-4}$ M).

If we assume that the rate of reaction between the peptide and the active products is limited by diffusion, the lifetime of the active product would be of the order of 10^{-5} seconds. The active particle is probably a triplet state of the dyes, while generation of radicals from the peptide results either from a dehydration or a charge transfer. It is noteworthy in this connection that the 2, 6-disodium salt of anthraquinone sulfonic acid — a typical dehydrating dye — similarly causes a slight chemiluminescence of the peptide on irradiation within the absorption band of the dye. On the other hand, participation of singlet oxygen in the process cannot be ruled out. The other parameters of sensitized chemiluminescence are typical of this peptide, although there is no radiation-induced destruction of the free radicals. Thus, in this case also the main characteristics of the process agree quantitatively with the schemes proposed above.

Another study [15] deals with chemiluminescence of glycyl-tryptophan and HSA solutions exposed to a silent electric discharge in argon and oxygen. The results of this study deserve attention in connection with the radiobiological effect of relatively slow electrons on biochemically important compounds. It was shown that chemiluminescence results from an indirect effect of the irradiation. Another finding reported in the same publication is that ozone likewise induces chemiluminescence. Kinetic analysis shows that the other parameters of chemiluminescence coincide with those determined in the case of UV excitation.

A survey of works published on glycyl-tryptophan solutions leads to the conclusion that the most important unsolved problems are the chemical mechanism of the processes and the nature of the emitter. For such a complex molecule there are obviously a number of possible pathways for photo- and radiochemical transformations. Studies of the photodestruction of tryptophan [90—93] show that a moderate degree of transformation gives rise to the following major products: formyl kynurenine, kynurenine, anthranilic acid (reactions involving a rupture of the pyrrole ring), indole acetic acid, alanine, and ammonia. Suitable emission levels exist in ortho-substituted aromatic ketones of the kynurenine type, which are formed only in the presence of oxygen (chemiluminescence is likewise observed only in the presence of oxygen). It was assumed on these grounds that the emitter (chemiluminescence center) is kynurenine — a possible product of disproportionation of free radicals [46, 50].

This assumption was confirmed by Sapezhinskii et al. in a study [94] of the relationship between intensity of chemiluminescence and accumulation of kynurenine, chemiluminescence and luminescence spectra of photooxidation products, kinetics of luminescence and accumulation of products. It was found that kynurenine accumulates slowly at first until about 1 minute after the start of irradiation, when the concentration of this ketone begins to rise linearly; the intensity of chemiluminescence rises at first to a peak and then steadily declines (consumption of the peptide). These values were found to be quantitatively interrelated. The corresponding equations were obtained on the basis of the general scheme of photochemiluminescence of the peptide, and the chemiluminescence yield per molecule of ketone formed was calculated (approximately $0.7 \cdot 10^{-6}$). Addition of cysteine to the system sharply reduces the intensity of chemiluminescence although the yield remains almost unchanged. It was concluded from these experiments that the excited product may be a kynurenine molecule in an excited-electron state. In addition to the comparative kinetic analysis, a study was made of the spectral characteristics of chemiluminescence and luminescence of photooxidation products of the peptide. The luminescence spectrum (after extraction of anthranilic acid and introduction of other corrections) resembles the chemiluminescence spectrum.

Finally, it was found that the kinetics of the accumulation of kynurenine determined from absorption and from luminescence (low optical density) are identical. Thus, the kinetic and spectral data indicate that the emitter of photochemiluminescence of glycyl-tryptophan solutions is a molecule of an aromatic ketone of the kynurenine type. The active compound appears to be formyl kynurenine rather than kynurenine, but a definitive conclusion requires further tests. The levels and transitions responsible for emission of light remain unknown (there are practically no data on this topic). Preliminary findings indicate that kynurenine-type groups serve as emitters in photochemiluminescence of proteins also.

Recently the effect of certain organic solvents on the kinetic curves of afterglow of glycyl-tryptophan has been studied. The purpose of these investigations was to determine the manner in which polarity of the surroundings influences the chemiluminescence processes, since the differences between chemiluminescence of proteins and that of the peptide may be due to more hydrophobic environs of the tryptophan residues in proteins. It was found that introduction of methanol, acetone and dioxan lowers the intensity of chemiluminescence and slows down the decay of afterglow; the kinetic curves of afterglow are not exponential under such conditions.

The effect of organic solvents is reversible. The introduction of these compounds lowers the intensity of glow and slows down the course of the kinetic curves. If water (buffer) is added after some time the intensity of chemiluminescence rises sharply and the curves revert to their original form. These findings prove the solvation nature of the effect of such solvents. The initial intensity of chemiluminescence, the light sum, the time required for the glow to decay by a factor of e, and the parameter of the form of the kinetic curves were calculated on the basis of the kinetic curves of afterglow; the effect of concentration of the additives on these parameters was also determined.

The results obtained were compared with data derived from a calculation of the probable mechanisms of the process. One of the schemes was found to agree with all the available experimental data. In an aqueous solution, the peroxide free radicals form hydrogen bonds with water molecules [95–97] and preferentially isomerize into secondary radicals. In organic solvents there is a shift of the equilibrium (owing to lack of water) and the desolvated radicals disproportionate. Comparison of the theoretical results with experimental data confirms this scheme. It became clear that intensity of chemiluminescence decreases with the dielectric constant of the medium.

Thus, the change in form of the kinetic curves of afterglow on the introduction of organic solvents may be attributed to a specific solvation of the peroxide radicals from the peptide with water molecules. The decrease in intensity and yield of chemiluminescence as the dielectric constant diminishes reflect universal properties of the medium as a dielectric and apparently results from Coulomb interactions. The divergences between parameters of the induced chemiluminescence of proteins and those of glycyl-tryptophan can be explained in terms of solvation interactions on the basis of these data. The chemiluminescence method possibly has a future application in quantitative analysis of the properties of tryptophyl residues in proteins.

THE EFFECT OF VARIOUS COMPOUNDS ON THE CHEMILUMINESCENCE OF PROTEIN SOLUTIONS

The effect of various additives on prolonged afterglow of proteins and peptides deserves attention for several reasons. To begin with, the indirect effects of radiation predominate in dilute aqueous solutions of proteins and peptides. The addition of various acceptors of the active products of the radiolysis of water must therefore lead to a decrease in both intensity and ionization yield of induced chemiluminescence. Using competitive acceptors it is possible to determine the relative rate constants of the reactions between the acceptors and the active products of radiolysis of water. Partial results of such a determination have been published [16], but a thorough study has yet to be made; the chemiluminescence method has so far not been applied to this question.

Data relating to the influence of the various metabolites present in natural biological systems on the radiation-induced lesions of proteins are of particular interest. Not only the reactions and processes involved in formation of free radicals from proteins are relevant here but also the chemical reactivity of these radicals and the pathways leading to manifestation of the lesion either within the protein macromolecule or in reactions with neighboring compounds. It is also important to determine the mechanism of the radioprotective effect of various compounds and to quantitatively evaluate the reaction between radioprotectors and free radicals formed from the proteins. A number of studies [33, 41, 44] have been devoted to these topics. The additives are usually introduced after irradiation so that primary processes need not be taken into account. About 100 compounds of various types have been tested so far, and their effect on chemiluminescence of proteins expressed in semiquantitative or quantitative terms.

One of the studies cited [44] examines the effect of 70 different compounds — amino acids, peptides, sugars, alcohols, aldehydes, ketones, phenols and variable-valence metal ions — on photochemiluminescence of BSA solutions at a constant concentration of the additive $(5 \cdot 10^{-3} \text{ M})$. The effect of these compounds was expressed as the ratio between the rate constants of chemiluminescence decay in the presence and absence of the additive. If the ratio equals unity there is no effect; if it exceeds unity, the radicals react with the additive.

The effect of natural amino acids and a number of peptides on chemiluminescence was also determined. The effect of cysteine is most pronounced; valine, methionine, alanine, histidine and serine produce results exceeding the limits of experimental variability. The other amino acids practically do not affect the glow. Peptides exert only a minor effect. A study of the influence of sugars, alcohols, aldehydes and ketones showed that only saccharose and hexyl alcohol exert any influence on the afterglow. The effect of various phenols on this process was also determined. Some of these compounds react quite effectively with the protein radicals (propyl gallate, certain screened phenols). Copper and iron ions cause a slight increase in the rate constants. The studies by Sapezhinskii et al. [33, 34] contain data on the effect of various compounds tested as radioprotectors according to the data available.

TABLE 3.6. Rate constants of reactions between protein radicals and radioprotectors

Compound	Rate constant, $M^{-1} sec^{-1}$	Compound	Rate constant, $M^{-1} sec^{-1}$
Cysteamine	4.6	Mercaptamine disulfide	0.4
Thiourea	2.9	2-Propyl-6-methyl-3-	
Cysteine	2.6	oxypyridine	0.3
Aminoethylisothiouronium		Aniline	0.2
hydrobromide (AET)	1.7	Serotonin	0.2
Aminopropylisothiouronium		Ascorbic acid	0.1
hydrobromide (APT)	1.6	5-Methoxytryptamine	0.0
Glutathione	1.3	Glucose	0.0
Propyl gallate	1.2	Hydroxylamine	0.0
4-Hydroxy-3,5-di-tert-		Sodium nitrite	0.0
butylphenylethylamine	0.6		
Sodium sulfite	0.6		
Sodium thiosulfate	0.6		

The rate constants given in Table 3.6 are specific values which do not depend on the concentration of the compound introduced. It is difficult to compare them with the results of experiments on animals in vivo because of the lack of comparative quantitative parameters for a sufficiently large range of compounds.

It is evident from the available data that compounds containing sulfhydryl groups, isothiuronium compounds and phenols react effectively with the free radicals of the protein. The protective effect of serotonin and 5-methoxytryptamine apparently involves other mechanisms, or alternatively these compounds undergo substantial transformation in the course of metabolism. Thus, many compounds react readily and highly selectively with the free radicals of proteins. The causes of this specificity remain a mystery and require further research.

A very important question, from the radiobiological viewpoint, is the possibility of a transfer of the free valence between the dissolved protein molecules. Such a chain reaction could spread the damage and increase destruction of cell components. We investigated this problem using HSA solutions, both native and stained with chemically bound fluoresceine isocyanate. As noted above, addition of xanthene dyes markedly intensifies the glow. Thus, the chemiluminescence yield of the protein chemically bound with the dye is 10−20 times greater than that of the native protein; the other properties are identical. The idea behind these experiments is quite simple:

the native protein is irradiated, and chemiluminescence appears; then the stained protein is added to the solution. If a valence transfer does take place the chemiluminescence intensity should increase in comparison with the control (to which native protein was added); conversely, addition of native protein to a previously irradiated solution of stained protein should decrease chemiluminescence intensity. These experiments were performed under various conditions of protein concentration, temperature, pH, etc. No increase or decrease in chemiluminescence was detected, which means that there is no intermolecular valence transfer as chain reaction in aqueous solutions of proteins (serum albumin). The reason for this may be that under our experimental conditions the protein molecules bear a negative charge which impedes their interaction; hydration effects may also be involved.

CHEMILUMINESCENCE OF COMPLEX OBJECTS

In this chapter we shall not consider photosynthetic chemiluminescence and other types of weak glow associated with photosynthetic pigments. Only a few studies of chemiluminescence in animal and plant cells and tissue and organ homogenates have been made so far. Of particular interes is the study by Konev et al. on photochemiluminescence of various strains of yeast and of homogenates of frog muscle and liver tissue [57]. These experiments, involving changes in the excitation wavelength, indicated that cell proteins are responsible for chemiluminescence [57]. It was shown that the glow in the cell is quenched in comparison with that observed in protein solutions, apparently as a result of intermolecular reactions in particular interactions between proteins and lipids.

Changes in the physiological state of the cells under the influence of temperature, UV radiation or the pH of the medium cause a flare-up of the glow associated with disturbances at the level of protein-lipid interaction. A slight afterglow induced by X-rays and gamma irradiation has been reported [21–24]. A quantitative analysis of these phenomena has not yet been made.

We shall discuss in greater detail the study by Dontsova et al. [58] on chemiluminescence of human blood plasma — a complete system comprising scores of components. The following questions were posed in this study. Which components of the plasma are responsible for photochemiluminescence? What is the contribution of the different fractions to this effect? Does blood plasma contain compounds which affect the yield of glow (activators, quenching agents)? Does plasma contain radical acceptors or sensitizers?

The experiments were carried out with lyophilized blood plasma and the different fractions of this. The protein composition of the plasma was determined by electrophoresis. The blood plasma contains both high and low molecular weight compounds. The category of high molecular weight compounds consists essentially of proteins (the concentration of nucleic acids and polysaccharides being very low). Dialysis showed that proteins

contribute more than 80% of the glow (in terms of the light sum). If proteins are responsible for the glow, it may be assumed that a glow comparable to that of native plasma can be obtained in a model solution of proteins
equivalent to the plasma in composition. Such a model solution was pre-
pared with serum albumin, α-, β- and γ-globulins at a concentration equivalent to 1.5% plasma. The parameters of photochemiluminescence of the
plasma and of the model solution were found to be similar. Analogous experiments were performed to determine the contribution of each protein to
the glow. It was found that the bulk of the glow comes from serum albumin
(77%); the contribution of the other proteins is as follows: α-globulin 7.9%,
β-globulin 8.4%, γ-globulin 3%, fibrinogen 1.5%.

In another series of tests, high concentrations of plasma were added to
irradiated HSA solutions in order to establish whether plasma contains
compounds influencing the yield of the glow. It was found that plasma does
not affect the afterglow of albumin. This indicates that the plasma samples
examined lack quenching agents, activators and acceptors of free protein
radicals. Kinetic measurements similarly showed that the main characteristics of photochemiluminescence of plasma differ only slightly from those
of albumin. The introduction of cysteine and propyl gallate into the irradiated solution accelerates decay of the glow, whereas eosin increases intensity of glow.

On the basis of these experiments, it seems likely that chemiluminescence
of more complex systems, such as homogenates or cells, will be found to
originate from only one or a few proteins, analogous to albumin in the case
of plasma. Little can be said at present about the mechanisms of chemiluminescence in the cell, except that this apparently contains a large assortment of compounds which influence chemiluminescence (radical acceptors,
activators, quenching agents, sensitizers) and greatly complicate these
effects. Ascites cells emit a gamma chemiluminescence which is weaker
by one order of magnitude than that of the proteins examined [98]. Further
research in this field will undoubtedly be of great interest.

THE METHOD OF INDUCED CHEMILUMINESCENCE
AS A SOURCE OF INFORMATION IN THE STUDY
OF PRIMARY RADIOBIOLOGICAL PROCESSES

From time to time articles stressing the importance of low-level glows
in terms of particular biological functions are published. In our view, such
hypotheses are erroneous and devoid of experimental support. Induced
chemiluminescence certainly deserves attention in its own right, like any
other new phenomenon, but the negligible yield precludes it from performing
any chemical or biological function. However, it does contain hidden information on the destructive processes occurring in proteins and other
objects.

Before discussing the possibilities of this method we shall briefly outline
its limitations on the basis of the data presented above. The main information derived from analysis of induced chemiluminescence is kinetic in nature.
Accordingly, reliability of the results depends on the thoroughness of the
kinetic determination of all the parameters of chemiluminescence under

various experimental conditions and establishment of a correlation between chemiluminescence and chemical transformations provoked by irradiation. The emission spectra provide data on the nature of the emitter, but they have not yet been obtained with sufficient accuracy because of the low intensity involved.

Chemiluminescence can be attributed largely to processes associated with the oxidative radiolysis and photolysis of tryptophan, either as an amino acid or included in peptides or proteins. This is so apparently because the chemiluminescence yield in the formation of kynurenine exceeds that obtained from other free-radical processes of biochemical importance. These considerations explain the limitations of the method.

Now we may turn to the possible applications of the method. First of all, a comment on its sensitivity is necessary in view of the fact that in conventional photometric devices the signal is equal to the noise at a luminous flux of the order of 10^2-10^3 quanta/(ml·sec). In the case of glycyl-tryptophan the afterglow effects can therefore be recorded at a dose rate of more than $0.1-1$ rad/sec and at doses exceeding $30-50$ rad. The lowest detectable breakdown rate of the peptide is 10^{-11} M/sec, while the lowest detectable concentration of peroxide-type free radicals is of the order of 10^{-9} M (10^{12} radicals/ml). Clearly, therefore, this method is far more sensitive than any other physicochemical procedure used for analysis of free-radical transformations. In other words, induced chemiluminescenc is a unique tool for the study of the photo- and radiochemical conversions of proteins within the range of so-called biological doses, when the various secondary and side effects are negligible.

The value of chemiluminescence in the analysis of primary radiobiological mechanisms can be assessed by considering the conventional scheme of radiation-induced events and selecting processes where application of this method appears justified.

I. Irradiation induces formation of active products of water radiolysis, which react with proteins to yield the free radicals of these. In this case the luminescent protein signals its own lesion. The following data can be obtained in this manner.

1. The nature of the active radiolytic products of water participating in formation of radicals from the biopolymer. Least well known are reactions involving excited water and the positive water ion.

2. Chemiluminescence can be used to determine the rate constants of reactions between active radiolytic products of water and molecules of proteins or other compounds. Apart from the overall rate constant, this method can also show the partial rate constant, i.e., the rate constant of the reaction between tryptophyl residues in the protein molecule and the radiolytic products of water.

3. Protection and sensitization with respect to radiation (notably the oxygen effect) can also be determined by this sensitive method at the level of the reactions of radioprotectors and radiosensitizers with the active radiolytic products of water.

II. The free radicals generated from biopolymers by reactions with radiolytic products of water or by other mechanisms can enter into various reactions and transformations leading to destruction. Chemiluminescence provides quantitative data on the yield, stationary concentration and destuction of these free radicals. For a number of reasons (solvation, the specific chemical nature of the processes), the free-radical processes in proteins last for a relatively long time — hundreds of seconds. Prolonged destructive processes also take place in animal and plant cells. These findings suggest that primary chemical processes in the cell are not very fast and do not end as soon as irradiation ceases. Although still difficult to assess from a radiobiochemical viewpoint, these prolonged processes must be considered in any analysis of radiation damage.

Generation of free radicals in the protein macromolecule may also be followed by various further transformations leading to decomposition of other amino acid residues, notably cysteine and cystine. Always an important event, the breakdown of tryptophan can even be fatal in some proteins. With small radiation doses, conventional physicochemical procedures cannot detect any conformational changes, even though the primary structure is disturbed (modified). It appears that chemical changes in the primary structure of proteins alter their tertiary structure, causing inactivation of enzymes even if tryptophan is not a constituent of the active center. The chemiluminescence method can be used in studying intramolecular reactions of free radicals of proteins or, more precisely, the most probable pathways leading to manifestation of lesions "inside" the macromolecule.

III. The peroxide-type free radicals of proteins can participate in numerous reactions with the surrounding molecules. Two aspects deserve attention here. Firstly, the lesion materializes by further intermolecular reactions between free radicals of the protein and various metabolites, both low and high molecular weight. Of particular interest are reactions of protein radicals with lipids, nucleic acids and polysaccharides. These reactions can take place both in solutions and at the level of the corresponding complexes. Induced chemiluminescence provides quantitative data on the rate constants of the reactions and thus points to possible ways leading to manifestation of the lesions at the level of subcellular structures. Secondly, this topic leads to the question of protection, which may be evident in such cases. As shown by the experimental evidence discussed above, radioprotectors react very effectively with free radicals of the protein. The mechanism of these reactions appears to be complex and composed of a number of intermediate stages, the identification of which will contribute much toward the understanding of the radioprotective action of many compounds. Induced chemiluminescence can be used in the construction of various tests for selecting effective radioprotectors under conditions simulating real intracellular processes. Finally, induced chemiluminescence can be applied in radiobiological dosimetry.

Even such a brief outline of the possible applications of induced chemiluminescence is enough to show the great value of this method to radiation biology.

CONCLUSION

Radiation-induced chemiluminescence is the most sensitive kinetic method for analyzing the free-radical transformations of proteins in aqueous solutions. The equipment required is quite simple. This method yields quantitative data (absolute or relative) on the following parameters: 1) rate of radiolysis (photolysis); 2) concentration of free radicals; 3) rate constant of the reaction of active products of water radiolysis (or excited states in the case of sensitized photochemiluminescence) with the molecules; 4) peroxide free radicals of proteins; 5) rate constants of reactions of these radicals with various compounds (metabolites, radioprotectors); 6) intramolecular free-radical transformations in macromolecules.

All these data can be obtained at low radiation doses (50−1,000 rad) and at low dose rates (5−10 rad/sec). Further research into the mechanism of induced chemiluminescence will extend the applicability of this method still further.

BIBLIOGRAPHY

1. Ahnstrom, G. and G.V.Ehrenstein.− Acta Chem. Scand. 13 (1959), 855.
2. Strehler, B.L. and W.A.Arnold.− J. Gen. Physiol. 34 (1951), 809.
3. Lelievre, B. and J.P.Adloft.− J. Phys., Paris 25 (1964), 789.
4. Ahnstrom, G.− Acta Chem. Scand. 15 (1961), 463.
5. Westermark, T. et al.− Ark. Kemi 17 (1961), 151.
6. Sapezhinskii, I.I. et al.− In: "Biolyuminestsentsiya," p.102. Moscow, "Nauka," 1965.
7. Sapezhinskii, I.I. et al.− Doklady AN SSSR 151 (1963), 584.
8. Sapezhinskii, I.I. and Yu.V.Silaev.− In: Svobodnoradikal'nye protsessy v biologicheskikh sistemakh," p.78. Moscow, "Nauka." 1966.
9. Sapezhinskii, I.I. and Yu.V.Silaev.− In: "Bioenergetika i biologicheskaya spektrofotometriya," p.4. Moscow, "Nauka." 1967.
10. Emanuel', N.M. et al.− Izv. AN SSSR, biol. ser., No.2 (1966), 183.
11. Höfert, M. Versuche über strahleninduzierte Chemilumineszenz.− Dissertation. München, 1964.
12. Höfert, M.− Angew. Chem. 76 (1964), 826.
13. Höfert, M.− Biophysik 2 (1969), 166.
14. Silaev, Yu.V. et al.− Khimiya Vysokikh Energii 2 (1968), 338.
15. Sapezhinskii, I.I.− Report Sympos. of the Moscow Society of Naturalists. In: "Sverkhslabye svecheniya v biologii," p.52. Moscow, MGU, 1969. (Russian)
16. Silaev, Yu.V. and I.I.Sapezhinskii.− Ibid., p.56.
17. Sapezhinskii, I.I. et al.− Doklady AN SSSR 159 (1964), 1378.
18. Emanuel', N.M. et al.− Izv. AN SSSR, techn. sci. sec., No.6 (1963), 1143.
19. Zhizhina, G.P. et al.− Doklady AN SSSR 163 (1965), 931.
20. Emanuel', N.M. et al.− Trudy Mosk. Obshch. Ispyt. Prir. 21 (1965), 119.
21. Avakyan, Ts.M. et al.− Trudy Mosk. Obshch. Ispyt. Prir. 21 (1965), 60.
22. Avakyan, Ts.M. et al.− Report Sympos. of the Moscow Society of Naturalists. In: "Sverkhslabye svecheniya v biologii," p.3. Moscow, MGU. 1969. (Russian)
23. Polivoda, A.I. et al.− Meditsinskaya Radiol., No.10 (1961), 90.
24. Tarusov, B.N. et al.− Radiobiologiya 1 (1961), 150.
25. Vladimirov, Yu.A. and F.F.Litvin.− Biofizika 4 (1959), 601.
26. Konev, S.V. and M.A.Katibnikov.− Biofizika 6 (1961), 638.
27. Katibnikov, M.A. and S.V.Konev.− Biofizika 7 (1962), 150.
28. Konev, S.V. et al.− Izv. AN BSSR, biol. ser., No.1 (1964), 76.
29. Tumerman, L.A. et al.− Biofizika 7 (1962), 21.
30. Kemp, T.J. and C.G.Clayton. Light-Induced Chemiluminescence, Isotope Research Division.− Wantage, Berks. 1964.
31. Stauff, J. et al.− Ber. Bunsenges. Phys. Chem. 70 (1966), 759.
32. Stauff, J. and D.Balzer.− Kolloidzeitschr. Z. Polymere 216−217 (1967), 376.
33. Sapezhinskii, I.I. and E.G.Dontsova.− Biofizika 12 (1967), 794.

34. Sapezhinskii, I.I. and V.A.Voloshin.— Biofizika, 13 (1968), 517.
35. Sapezhinskii, I.I. and E.G.Dontsova.— Khimiya Vysokikh Energii 4 (1970), 77.
36. Sapezhinskii, I.I. and E.G.Dontsova.— Biofizika 15 (1970), 745.
37. Herberg, R.— Science 128 (1958), 199.
38. Stauff, J. and H.Wolf.— Z. Naturf. 19b (1964), 87.
39. Hastings, J. et al.— Proc. Natn. Acad. Sci. U.S.A. 52 (1964), 1529.
40. Sapezhinskii, I.I. and N.M.Emanuel'— In: "Biolyuminestsentsiya," p.122. Moscow, "Nauka." 1965.
41. Sapezhinskii, I.I. et al.— Biofizika 10 (1965), 429.
42. Gibson, Q.H. et al.— Proc. Natn. Acad. Sci. U.S.A. 53 (1965), 187.
43. Hastings, J.W. et al.— Photoch. Photobiol. 4 (1965), 1227.
44. Sapezhinskii, I.I. and E.G.Dontsova.— Radiobiologiya 6 (1966), 646.
45. Sapezhinskii, I.I. et al.— Biofizika 11 (1966), 427.
46. Sapezhinskii, I.I.— Doklady AN SSSR 175 (1967), 1167.
47. Sapezhinskii, I.I. and Yu.V.Silaev.— Biofizika 12 (1967), 38.
48. Sapezhinskii, I.I. et al.— In: "Bioenergetika i biologicheskaya spektrofotometriya," p.26. Moscow, "Nauka." 1967.
49. Nisenbaum, G.D.— In: "Problemy sovremennoi biologii," p.99. Minsk, AN BSSR. 1967.
50. Sapezhinskii, I.I.— Biofizika 13 (1968), 720.
51. Aksentsev, S.L. et al.— Biofizika 13 (1968), 428.
52. Nisenbaum, G.D. et al.— Ibid., p.138.
53. Sapezhinskii, I.I.— Khimiya Vysokikh Energii 3 (1969), 325.
54. Aksentsev, S.L. et al.— Molekulyarnaya Biologiya 4, No.2. 1970.
55. Nisenbaum, G.D. et al.— Biofizika 14, No.3 (1969), 402.
56. Aksentsev, S.L. et al.— Report Sympos. of the Moscow Society of Naturalists. In: Sverkhslabye svecheniya v biologii," p.6., Moscow, MGU, 1969. (Russian)
57. Konev, S.V. et al.— Biofizika 16, No.1 (1970), 19.
58. Dontsova, E.G. et al.— Biofizika 13 (1968), 1061.
59. Krasnovskii, A.A. and F.F.Litvin.— Biofizika 13 (1968), 146.
60. Grossweiner, L.L. and A.F.Radde.— J. Phys. Chem. 72 (1968), 756.
61. Kuschnir, K. and T.Kuwana.— Chem. Commun., No.5 (1969), 193.
62. Krasnovskii, A.A. and F.F.Litvin.— Report Conf. on Photochemistry of Solutions, p.58. Kiev, "Naukova Dumka," 1969. (Russian)
63. Litvin, F.F. et al.— Usp. Fiz. Nauk 71 (1960), 149.
64. Goldheer, J.C. and G.R.Vegt.— Nature 193 (1962), 875.
65. Krasnovskii, A.A. and F.F.Litvin.— Report Conf. In: "Molekulyarnaya biofizika," p.174. Pushchino-na-Oke, 1966. (Russian)
66. Krasnovskii, A.A. and F.F.Litvin.— Molekulyarnaya Biologiya 1 (1967), 699.
67. Litvin, F.F. and A.A.Krasnovskii.— Doklady AN SSSR 173 (1967), 451.
68. Shiryaev, V.M. et al.— Report Sympos. of the Moscow Society of Naturalists. In: Sverkhslabye svecheniya v biologii," p.68, Moscow, MGU, 1969. (Russian)
69. Sapezhinskii, I.I. and E.G.Dontsova.— Ibid., p.53.
70. Streler, B.L.— Proc. 5th Intern. Biochem. Congress, p.85. Moscow, AN SSSR, 1961. (Russian)
71. Litvin, F.F. and V.A.Shuvalov.— Byull. Mosk. Obshch. Ispyt. Prir. 71 (1966), 141.
72. Litvin, F.F. et al.— Doklady AN SSSR 168 (1966), 1195.
73. Litvin, F.F. et al.— Trudy Mosk. Obshch. Ispyt. Prir. 16 (1966), 261.
74. Sapezhinskii, I.I.— In: "Svobodnoradikal'nye sostoyaniya v biologicheskikh sistemakh," p.25. Moscow, "Nauka." 1966.
75. Hastings, J.W. and G.Weber.— J. Opt. Soc. Am. 53 (1963), 1410.
76. Vasil'ev, R.F. and G.F.Fedorova.— Optika i Spektroskopiya 24 (1968), 419.
77. Parker, S.A.— Proc. R. Soc. A220 (1953), 104.
78. Hatchard, C.G. and C.A.Parker.— Proc. R. Soc. A235 (1956), 518.
79. Wagner, E.E. and A.W.Adamson.— J. Am. Chem. Soc. 88 (1966), 394.
80. Vasil'ev, R.F.— In: "Biolyuminestsentsiya," p.170. Moscow, "Nauka." 1965.
81. Silaev, Yu.V. et al.— Report Sympos. In: "Sverkhslabye svecheniya v biologii," p.56. Moscow, MGU. 1969.
82. Sapezhinskii, I.I. and E.G.Dontsova.— In: "Svobodnoradikal'nye protsessy v biologicheskikh sistemakh," p.50. Moscow, "Nauka." 1966.

83. U s a t y i , A.F. and Yu.S.L a z u r k i n.– In: "Elementarnye protsessy khimii vysokikh energii." Moscow, "Nauka." 1965.

84. V l a d i m i r o v , Yu.A. and D.I.Ro s h u p k i n.– Biofizika 9 (1964), 282.

85. V a s i l ' e v , R.F. et al.– Doklady AN SSSR 149 (1963), 124.

86. S a p e z h i n s k i i , I.I. and A.V.K u z n e t s o v a.– Biofizika 15 (1970), 531.

87. N i s e n b a u m , G.D.– Thesis. Minsk, 1969.

88. G r a n o v s k a y a , M.L. et al.– Radiobiologiya 5 (1965), 633.

89. A n b a r , M. and P.Ne t a.– Int. J. Appl. Radiat. Isotopes 18 (1967), 493.

90. M a t s u d a , G.– Nagasaki Igakkai Zassi 28 (1953), 438.

91. I o s h i d a , Z. and M.K a t o.– J. Am. Chem. Soc. 76 (1954), 311.

92. L e a v e r , I.H. and F.G.L e n n o x.– Photoch. Photobiol. 4 (1965), 491.

93. V l a d i m i r o v , Yu.A. and N.I.P e r a s s e.– Biofizika 11 (1966), 578.

94. S a p e z h i n s k i i , I.I. and E.G.D o n t s o v a.– Doklady AN SSSR 192 (1970), 129.

95. K n o p p e , D.G. and N.M.E m a n u e l '.– Uspekhi Khimii 24 (1955), 275.

96. A n d r o n o v , L.M. et al.– Doklady AN SSSR 173 (1967), 859.

97. A n d r o n o v , L.M. et al.– Doklady AN SSSR 174 (1967), 127.

98. K h a r a d u r o v , S.V. et al.– Report Sympos. on the Free-Radical State in Radiation Injury and Tumor Growth (Neoplasia), p.93. Moscow, Inst. Khim. Fiz. AN SSSR. 1970. (Russian)

Chapter 4

DISTURBANCE OF DNA FUNCTIONS AT THE INITIAL
STAGES OF RADIATION-INDUCED CELL DAMAGE

INTRODUCTION

The fundamental manifestation of cell life is protein synthesis. It is known that the process of protein biosynthesis, including replication of genetic material (chromatin), consists of several stages involving different mechanisms: synthesis of template RNA, synthesis of acceptor RNA molecules serving as amino acid carriers, formation of ribosomes and polyribosomes, activation of amino acids, etc. Research into these mechanisms is a leading trend in modern biology, success in this field enabling progress in biology as a whole and in particular in radiobiology.

The role of RNA in the transfer of hereditary information from DNA to the site of protein synthesis has been widely discussed in the literature during the last decade. Various types of cell RNA have been discovered and characterized, including both stable ones (ribosomal and transfer RNA) and a large category of unstable ones — the so-called informational or messenger RNA (mRNA). Their basic properties — metabolic behavior, sedimentation characteristics, DNA-like composition, etc., were determined.

Artificial hybridization experiments have demonstrated the complementary nature of the nucleotide sequence of nascent RNA with respect to the homologous DNA. At the same time, considerable progress has been made in the in vitro analysis of the structure and properties of DNA as a template for RNA synthesis. Various working hypotheses on the mechanism of RNA synthesis have been elaborated and the gene specificity of cell RNA has become an established fact. The study of all types of template RNA, their synthesis on DNA and their subsequent fate in the cell is therefore of great interest from the point of view of understanding the effect of ionizing radiations on the development and change of these processes. The effect of radiation on replication and transcription will be discussed separately in view of recent concepts of the functional organization of DNA.

THE EFFECT OF RADIATION ON REPLICATION
(DNA SYNTHESIS)

The living cell is extremely sensitive to ionizing radiations. More than 25 years ago Euler and Hevesy [1] found that X-ray irradiation suppresses

the incorporation of P^{32} into the DNA of the Jensen sarcoma. They interpreted this phenomenon as an inhibition of DNA synthesis in the tumor. It
has been established since that this original finding — a disturbance of DNA
synthesis — is a universal manifestation of the effect of radiation on living
cells.

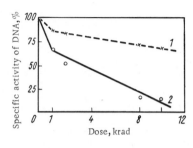

FIGURE 4.1. Effect of the dose and
type of radiation on the rate of DNA
synthesis in pea seedlings [8]:

1) gamma radiation; 2) neutron
radiation.

Hevesy made another interesting observation at the time — namely, that
once the process has been inhibited by 45—50%, further change requires a
much greater dose of irradiation [2]. On the basis of this finding, a number of
scientists tried to determine inhibition of the incorporation of the precursor
into the DNA of various organs and tissues as a function of the radiation
dose. It became clear from these experiments that the effect of irradiation
on DNA is immediate and further changes do not take place for a considerable period [3, 4]. Similar data published by other authors are given in the
review by Holmes [5]. Using modern methods, Vainson [6, 7] made a
thorough study of the inhibition of DNA synthesis in a synchronous culture of
HeLa cells subjected to total and local irradiation. He examined the dose
effect of the inhibition of synthesis under the influence of alpha particles
and gamma radiation. In both cases the dose-effect curve was found to
consist of two parts. The "radiosensitive" part differs from the "radioresistant" part by at least an order of magnitude in terms of the doses
involved. Figure 4.1 shows the inhibitory effect of gamma and neutron
radiations on DNA synthesis in pea germs [8].

It is evident from the literature that a radiosensitive stage of DNA
synthesis has been detected in a large variety of objects. Various data
have been used to explain the causes of this inhibitory effect of radiation
on DNA synthesis and the numerous working hypotheses drawn up have
contributed much toward a better understanding of the mechanisms of the
radiation effect. Figure 4.2 shows schematically the processes which may
be involved in the radiation-induced disturbance of DNA replication.
Causes of the sensitivity and resistance of the different parts of the dose
curve had to be determined in order to explain the existence of two stages
in inhibition of DNA synthesis.

It is an established fact that DNA synthesis is closely associated with
cell division. Howard and Pelc [9] examined the radiosensitivity of the
mitotic cycle in bean root tips. They showed for the first time that the
life cycle of the cell can be divided into several periods, which were

designated G_1, S, G_2 and M.   The dose required to inhibit incorporation of precursors into DNA by a certain amount during stage S is several times greater than the corresponding dose for stage G_1.   Similar results were obtained with regenerating liver, synchronous cultures of HeLa cells and human bone marrow [11, 12].

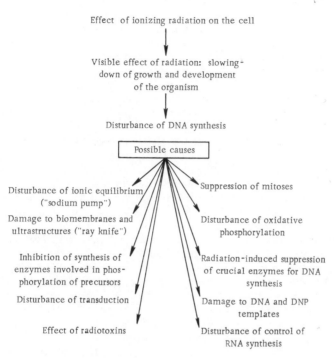

Effect of ionizing radiation on the cell

Visible effect of radiation: slowing-down of growth and development of the organism

Disturbance of DNA synthesis

Possible causes

Disturbance of ionic equilibrium ("sodium pump")

Damage to biomembranes and ultrastructures ("ray knife")

Inhibition of synthesis of enzymes involved in phosphorylation of precursors

Disturbance of transduction

Effect of radiotoxins

Suppression of mitoses

Disturbance of oxidative phosphorylation

Radiation-induced suppression of crucial enzymes for DNA synthesis

Damage to DNA and DNP templates

Disturbance of control of RNA synthesis

FIGURE 4.2. Possible causes of the disturbance of DNA synthesis under the influence of radiation

Discussing the possible relationship between inhibition of DNA synthesis and suppression of mitoses, Howard [9] notes that mitosis is more radiosensitive than DNA synthesis.   A delay in mitosis becomes evident at doses as low as 4—50 r.   The most radiosensitive stage is the early prophase, when DNA synthesis is already completed.   Hollaender [13] points out in this connection that a hypotonic solution causes a delay in mitosis.   This means that a simple physical process may be responsible for the suppression of mitosis.   All this evidence suggests that the latter phenomenon cannot be attributed to a delay in DNA synthesis although there is no doubt that such a delay prevents mitosis in cells irradiated during early interphase.

The effect of radiation on functional organization of the cell has been approached from various angles.   The destruction or transformation of the different nucleotides in DNA, the breakdown of sulfur-containing groups, the damage inflicted on ultrastructures and biomembranes ("ray knife"), and the lesions of critical enzymes involved in such processes as DNA synthesis or oxidative phosphorylation in the mitochondria and nucleus have all been studied.

A vast store of experimental evidence shows quite clearly that radiation-induced injury of the nucleus has more severe consequences for the cell than damage of the cytoplasm. It appears that the cytoplasm is capable of effecting partial repair in the presence of a normally functioning nucleus. The irradiation instantaneously damages various intracellular membranes, thus disturbing the permeability and ionic equilibrium [13]. This important type of radiation damage, however, is largely confined to ultrastructures in the cytoplasm, which is characterized by a multiple recurrence of components and identical macromolecules and microstructures, and is also distinguished by the lability and intensity of its various metabolic pathways and the presence of autonomous adaptive and repair functions. The ability of the cytoplasm to repair damage more rapidly can be attributed to these factors [14].

Some workers assumed that the drastic change in the rate of DNA synthesis at low radiation doses (the first branch of the dose-effect curve) results from inhibition of the oxidative phosphorylation of precursors in the nucleus, whereas the subsequent change in the rate of DNA synthesis can be attributed to disturbances of mitochondrial activity and changes in the DNA template [15, 16]. However, the concept that inhibition of DNA synthesis at low doses results from reduced activity of enzyme systems faces a serious objection in so far as in the initial part of the dose curve LD_{37} is slightly more than 1 krad while the complete arrest of the phosphorylation of the precursors requires doses of 10 krad or more. It is known, on the other hand, that nuclear phosphorylation ceases at 100 r [16], which would mean a virtual arrest of DNA synthesis in view of the limited reserve of precursors in the cell, whereas gamma irradiation does not inhibit DNA synthesis by more than 50% [2].

The decrease in oxidative phosphorylation in the mitochondria, determining the energy level of the cell, takes place as much as several hours after irradiation. Moreover, direct irradiation of isolated mitochondria requires very large doses in order to produce a comparable effect. Disturbance of oxidative phosphorylation in the nucleus is essential for the formation of polyphosphate derivatives of deoxyribonucleotides, but it is known that ATP can migrate directly from the mitochondria into the nucleus. Which of these processes is affected by ionizing radiations remains to be established [17]. It was thus clear that inhibition of DNA synthesis does not directly result from a deficiency of phosphorylated precursors — a conclusion confirmed by Libinzon [18].

Similarly, no changes were detected in the activity of DNA polymerase in irradiated tissues [15, 19]. A de novo incorporation of labeled deoxyribonucleotides was noted in radiation-damaged DNA. Owing to the radio resistance of DNA polymerase, it appears that any energetically suitable phosphorylated nucleotides are joined together on the template and that their level in the irradiated cell remains essentially unchanged [20, 21].

Thus, over the course of time, various explanations stressing different processes and structures were proposed in connection with the radiation death of the cell, but none of these hypotheses was generally accepted. It appears that the available procedures do not provide a reliable picture of the role and sequence of the different components of the cell with respect to radiation damage.

According to the current view, the postradiation synthesis of DNA depends mainly on two factors: radiotoxins and lesions of the DNA template [22]. These factors will be briefly discussed here. As long ago as 1959, Kuzin presented evidence shedding fresh light on the causes of radiation-induced inhibition of DNA synthesis and mitoses. When plants were locally exposed to gamma quanta, a long-range effect of the ionizing radiation on mitotic activity at the growth points was detected. The suppression of mitosis results from the formation of quinoid products of the enzymatic oxidation of tyrosine in the irradiated plant. These products are also present in normal plants and apparently control cell division. Removal of the irradiated part of the plant prevents inhibition of mitotic activity in the rootlets. The appearance of antimitotic agents depends on the dose and the time after irradiation. Radiotoxins are the object of increasing attention on the part of scientists. Many radiobiologists stress the importance of the newly formed toxic compounds in the development of radiation damage, especially in connection with inhibition of DNA synthesis [22].

The second factor is the effect of ionizing radiations on the structure of DNA. The chromosome structure consists essentially of supramolecular nucleoprotein formations. These are known to be highly radiosensitive. Conceivably, the high radiosensitivity of the cell is associated with disruption of the complex macromolecular organization or disordering of a supramolecular structure [23, 24]. However, structural disturbances in supramolecular DNA formations are not necessarily accompanied by a direct lesion of the DNA macromolecule [25].*

It is not yet known which of these structures (static or functional) is more vulnerable to irradiation. Apparently, both are susceptible. Numerous reviews deal with the effect of ionizing radiations on the DNA structure in vivo and in vitro. Radiation causes a variety of lesions in DNA in vitro. These include the radiolysis of bases, depolymerization, single and double breaks in DNA chains, configurational changes in DNA (loss of viscosity without a change in molecular weight) resulting from "compression" of DNA and local dissociation of DNA and protein.

Ionizing radiation causes three main types of lesions in DNA in the cell: loss of viscosity of the nucleoprotein and a decrease in the viscoelasticity of the supramolecular DNA complex, changes in the nucleotide composition of DNA, and enzymatic breakdown of DNA. Particular attention has been devoted to changes in the nucleotide composition of DNA as a result of irradiation. Conceivably, the presence of lesions in some of the nucleotides can impede or even rule out their pairing with the complementary partners during biosynthesis of new molecules, the result being a delayed or irregular synthesis of DNA, ultimately leading to mutations.

The most vulnerable sites of the DNA molecule are the A—T nucleotide pairs [25, 26]. Thymine apparently breaks down first, turning into a stable organic peroxide in the presence of oxygen. Under in vivo conditions this

---

* It has been shown [Struchkov, V.A. and N.B.Strazhevskaya.— Radiobiologiya, 8:787. 1968] that disturbance of the viscoelasticity of supramolecular DNA is correlated with single breaks.— The Editor.

compound is incapable of firm complementary base pairing, which leads to errors during replication and subsequent genetic anomalies. The radiolysis of nucleic acids and the associated problem of radiation mutagenesis have been studied in some detail by Scholes [26] and Shal'nov [27]. Little is known so far as regards the course (trend) of the lesion, i. e., whether "everything" begins from thymine or whether the trend is from the single breaks to the thymine.

All these results were obtained with total DNA preparations. In reality the extent of the damage differs from one cell to another within the irradiated organism. Using modern techniques, we were able to divide the DNA molecules in two major fractions according to the degree of their lesions. The first fraction consists of molecules bearing latent (potential) lesions but retaining their biological activity (that is, they are still capable of replication and transcription). The second fraction is composed of molecules with obvious structural damage manifested by a number of symptoms including a change in the nucleotide ratio. Fractionation of the DNA preparations has shown that the population of DNA molecules isolated from irradiated cells is rich in fragments containing mostly (up to 60%) G–C nucleotide pairs [28, 29]. These studies confirm our previous data on the lability of the A–T nucleotide pairs and the fact that the resulting change in the nucleotide composition of DNA stems from a radiation-induced lesion of thymine [30, 31].

Thus, analysis of existing data concerning the inhibitory effect of radiation on DNA synthesis (replication) leads to the following conclusions.

1. The disturbance of DNA synthesis is a universal manifestation of the effect of radiation on living cells.

2. The dose curve of the inhibition of DNA synthesis consists of two parts. The first or "radiosensitive" part covers a range of small doses which vary according to the object and the experimental conditions, and leads to inhibition of DNA synthesis by 30–50%. Greater effects can only be obtained by using very large radiation doses. The "radiosensitive" and "radioresistant" parts of the dose curve differ from one another by at least two orders of magnitude.

Various facts and working hypotheses have been put forward to explain the inhibitory effect of radiation on DNA synthesis (see Figure 4.2) with particular reference to impairment of the reproductive function of DNA; some of these theories still deserve attention.

Comparison of the effects of radiation on DNA in solution and in the living cell leads to the conclusion that the second or "radioresistant" part of the dose curve of inhibition of DNA synthesis can be attributed to various lesions in the structure of this compound.

A new interpretation of the difference in the radiosensitivity of the inhibition of DNA synthesis is emerging on the basis of recent concepts of the functional organization of DNA. The first (radiosensitive) stage of the inhibition of DNA synthesis possibly involves the indirect participation of radiotoxins generated by the radiation. However, other factors may also be responsible. In our view, the radiosensitive stage of the inhibition of DNA synthesis can be attributed to a disturbance of the mechanisms controlling RNA synthesis. This assumption finds support in the following data on the effect of radiation on the informational function of DNA (i. e., the transcription process).

RADIATION DAMAGE TO THE INFORMATIONAL FUNCTION OF DNA

Strictly speaking, the informational function of DNA consists of the following processes: activation of the template (inclusion of new genes); operation of DNA as a primer in the RNA polymerase system; the transcription process, i. e., the decoding of information (RNA synthesis proper); and control of the biosynthesis of protein.

Data available on the effect of radiation on the synthesis and metabolism of RNA are highly contradictory and have been erroneously interpreted as proof that RNA synthesis is radioresistant in comparison with DNA synthesis. This conception has arisen largely because the researchers have been unable to detect the changes in RNA synthesis; some of them have reported an inhibition of RNA synthesis, others have not observed this effect; moreover, an intensification of RNA synthesis at various times after the irradiation is stressed in some publications. This intensification has been frequently interpreted as a relative repair of initial damage in the cells. Below we shall discuss these works consecutively.

As long ago as 1947, Holmes found that irradiation inhibits the synthesis of both RNA and DNA in tumor tissue. However, inhibition of RNA synthesis was less pronounced than that of DNA [32]. Many authors found no appreciable changes in the biosynthesis of cell RNA under the influence of irradiation even when DNA synthesis was markedly inhibited. Thus, Mil'man and Shapiro [33] did not detect any changes in the RNA content in the 4-blastomere stage of loach embryos within 4 hours after X-ray irradiation at a dose of 40 kr. The effect of radiation-induced inactivation of the nuclei on the synthesis of mRNA was studied in the same object (loach embryos). Column chromatography with MAK (methylated albumin kieselguhr) showed that X-ray irradiation of the embryos at a dose of 50 kr completely inactivates DNA synthesis but does not inactivate RNA synthesis in the embryonic cells [34]. On the other hand, Kafiani and his colleagues [35] subsequently detected an inhibition of RNA synthesis during various embryonic stages of the loach.

Interesting comparative studies of the level and specific activity of RNA after short and long periods of time after irradiation have been published by a number of authors. Long after irradiation there is a marked intensification of RNA synthesis while the rate of DNA synthesis remains below normal [36, 37]. Dikovenko [38] determined incorporation of P^{32} into the nucleic acids of the spleen, thymus, bone marrow, small intestine mucosa and liver during the first two hours after X-ray irradiation at a dose of 2,000 r. There was a distinct decrease in the level of incorporation of the label in DNA. Similar results were obtained with respect to RNA in all organs except the liver. In the liver of irradiated rats there was a slight initial decrease in the rate of incorporation of the label into the nucleic acids followed by a progressive increase to a level indicating a normal rate of synthesis in the case of DNA and a rate far above the normal in the case of RNA [56].

Several publications deal with the adverse effect of radiation on the synthesis of nucleic acids in plants [39—41]. As in the animal organism, a disturbance of DNA synthesis and a very slight change in RNA synthesis are observed within two hours after irradiation. We tested the effect of gamma irradiation on the synthesis of DNA and RNA in 7-day old maize germs photosynthesizing in a $C^{14}O_2$ atmosphere. It was found that irradiation inhibits synthesis of DNA and RNA to a different extent [41].

Budnitskaya and her colleagues [42] examined the effect of irradiating wheat and barley seeds with gamma rays on the nucleic acid content of these plants during ontogeny. They found that an irradiation dose of 5, 10 or 40 kr disturbs the quantitative correlation between biosynthesis of nucleic acids and protein. Irradiation of the embryos in dry barley seeds causes decrease in the DNA content, while at the same time the amount of protein increases and the RNA level remains almost unchanged. Twenty-four hours after the beginning of germination the irradiated wheat seeds show inhibition of DNA synthesis and intensification of protein synthesis.

As noted above, the effect of radiation on cell RNA is detectable and sometimes pronounced. Fractionation of total preparations into nuclear and cytoplasmic RNA showed that ionizing radiations exert different effects on the various categories of cell RNA. Most workers agree that synthesis of nuclear RNA is inhibited while that of cytoplasmic RNA is enhanced. Payne and his colleagues [43] found inhibition of the incorporation of P^{32} into the DNA and nuclear RNA of rat and mouse liver 24 hours after general X-ray irradiation at a dose of 2.5 kr. The radiosensitivity of nuclear RNA was also stressed elsewhere [44], but no attempt was made in these studies to fractionate the ribonucleic acids.

X-ray irradiation of Ehrlich ascites carcinoma at a dose of 750—3,000 r inhibits incorporation of uracil-C^{14} into nuclear and nucleolar RNA by 70% but does not inhibit incorporation into cytoplasmic RNA. Fractionation of liver and thymus tissues revealed slowing down of the incorporation of P^{32} and adenine-C^{14} into the nuclear RNA fraction and simultaneous enhancement of the incorporation of P^{32} into the total RNA fraction [45, 46]. Budnitskaya [47], working in Errera's laboratory, showed that irradiation of HeLa cells (human tumor cells) at a dose of 100, 300 and 900 r rapidly slows down incorporation of cytidine-H^3 into nuclear and nucleolar RNA. With prolonged incubation of the cells (1—4 hours), however, this inhibitory effect is less pronounced.

Using the same cells, Kalendo [48] did not detect any appreciable inhibitory effect of irradiation on incorporation of uridine-H^3 by pulse labeling. Radioautography showed that the initial rate of RNA synthesis in the nucleus remained unchanged for a period of three hours after irradiation of the cell culture at a dose of 10 kr.

Cells of the differentiated adult organism have been classified on the basis of their reaction to irradiation into radiosensitive and radioresistant. Many workers have shown that irradiation slows down incorporation of precursors into nuclear RNA and lowers the RNA content in the radiosensitive organs of the hematopoietic and germinative systems and the gastrointestinal

tract.   On the other hand, radioresistant objects such as liver, kidney,
nerve and muscle tissues react differently to irradiation.

Hevesy [2] first reported that reduction in the DNA content of irradiated
rat liver is accompanied by a rise in RNA content.   Intensive research in
this field which started some 10 years later showed similar increases in
the RNA content of other irradiated organs and tissues besides liver cells
[49−55].   Cheika and Nosek[53] determined the nucleic acid content of the
spleen and liver of rabbits irradiated with a dose of 600 r in the case of total
irradiation and on screening the organ with a lead plate.   They found a drastic
reduction in the RNA content of the spleen (per 1 g of dry substance), and a
distinct and significant increase in the RNA content of the liver.   When the
organ is screened these changes are standardized.

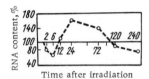

FIGURE 4.3.   Dynamics of changes
in the content of RNA in Purkinje's
cells of the cerebellum after
irradiation

In a histochemical study of the nucleic acid content of liver cells, Gubin
[52] found that gamma irradiation of rats at a dose of 800 r sharply lowers
the DNA content.   The DNA content remains lower than normal for 2−3 days
after irradiation, whereas the RNA content of the cytoplasm begins to increase on the second day.   Using UV cytospectrophotometry, V. I. Sharobaiko
[55] determined the amount of cytoplasmic RNA in Purkinje cells of the
cerebellum during the first 12 hours and for 10 days after the irradiation
of rats with a 400 r dose.   She detected phase-type fluctuations in the RNA
content of the nerve tissue.   Within 2−6 hours the RNA content decreased
by 20−32%.   After 12 hours it rose above the level found in nonirradiated
animals.   On the first and third days the RNA content was respectively 61
and 40% above normal.   The RNA content returned to normal by the fifth
day and dropped to 20% below the normal value by the tenth day (Figure 4.3).
Thus, the general dynamics of the RNA content of nervous tissue resemble
the corresponding changes observed in other tissues.   Similar changes in
the biosynthesis of RNA in the nervous tissue of rats were reported by
other workers [57].

The Radiobiological Laboratory of the Belgrade Institute of Nuclear
Sciences is engaged in a study of the effect of radiation on the composition
and functions of subcellular structures of cells and tissues, the DNP of rat
embryo liver, the structure of DNA, protein metabolism, the metabolism of
RNA in resting rat liver, etc.   Thus, Petrovic and his colleagues [58] examined the metabolism of RNA in different subcellular fractions of liver.
For this purpose they fractionated the nuclear and cytoplasmic RNA, but
did not identify the different fractions.   The synthesis of nuclear RNA was
found to be more radiosensitive than that of cytoplasmic RNA.   A change
in the nucleotide composition of nuclear RNA was detected as early as

4 hours after irradiation of the rats at a dose of 850 r.   Early changes were subsequently found in the RNA metabolism of resting liver cells following X-ray irradiation [59].   These changes were determined by various methods including tissue fractionation and RNA fractionation on MAK columns.   The RNA was labeled with the following precursors: orotic acid-6-C^{14} and adenine-8-C^{14}.   Conspicuous changes were detected both immediately and 30 minutes after irradiation: incorporation of the label into 35 S RNA and 4 S transfer RNA dropped to 60%.   At the same time incorporation of the label into the precursor of ribosomal RNA continued up to 3 hours after irradiation, and there was simultaneous accumulation of the label in the RNA fraction extracted from the microsomes.   The authors conclude that the metabolic effects of radiation begin in the nucleus and are observable in the cytoplasm only after a considerable time.

Petrovic and his colleagues also tested the effect of X-ray irradiation on RNA metabolism by ultracentrifugation in a sucrose gradient [60].   After the rats had been exposed to a lethal dose, the liver cells were fractionated into a number of subcellular fractions by differential centrifugation and the nucleotide composition of the nuclear and cytoplasmic RNA was determined in each fraction.   There was a marked difference in the specific activity of uridylic and cytidylic acids.   The ratio between uridylic and cytidylic acids rose sharply soon after irradiation, indicating slowing down of the synthesis of cytidylic acid in the nuclear fraction of liver RNA.

Hudnik-Plevnic [61] examined the effect of gamma irradiation on RNA metabolism in bacteria.   Experiments with pulse-labeled RNA revealed the following differences between the RNA of intact and irradiated cells: 1) incorporation of the label into the RNA of all fractions is enhanced in the irradiated bacteria; 2) increase in the level of incorporation varies from one fraction to another — it is much greater in 23—16 S RNA than in 14—10 S RNA.

This effect appears at a dose of 20 kr, i. e., there is a distinct dose-effect relationship (Figure 4.4).

Hudnik-Plevnic attributes these results to a radiation-induced derepression of the bacteria following irradiation.   Simic and co-workers [62] determined the synthesis and base composition of nascent RNA in the liver of 15-day old rat embryos after irradiation at a dose of 100 r on the 9th day of embryogeny.   The specific activity of RNA extracted from the irradiated embryos was roughly twice the normal value.   Centrifugation in a sucrose gradient showed increased activity of all fractions, especially the 35—40 S. The GC/AU specificity coefficient of the newly formed RNA in the liver of irradiated embryos was 1.93, compared with 1.31 in the controls.   This indicates activation of the synthesis of ribosomal RNA.

Mant'eva and her colleagues [46] observed activation of the biosynthesis of ribosomal and DNA-like nuclear RNA as well as ribosomal and transfer RNA in the cytoplasm of rat liver two hours after irradiation at a dose of 1–5 kr.   Similar data were reported elsewhere [63, 64].   By means of chromatography on MAK columns, these authors determined the specific activity of RNA fractions obtained by the hot phenol method from rat liver and spleen at the early stages of severe radiation damage.   Marked activation of the biosynthesis of mRNA was observed in the liver and pronounced inhibition in the spleen; these changes were most obvious two hours after irradiation and by 24 hours the rate of biosynthesis approached the normal level. No changes in the biosynthesis of other types of RNA were found under these conditions.

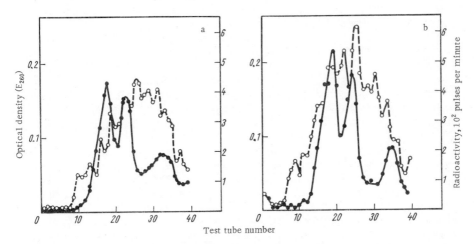

FIGURE 4.4. The effect of radiation on the synthesis of pulse-labeled RNA in bacteria [61]:

a) normal; b) after irradiation at a dose of 20 kr. Sedimentogram of a RNA preparation from Salmonella typhi-murium cells labeled with radioactive phosphorus. Sucrose gradient 4–20%, Mg^{2+} $5 \cdot 10^{-4}$ M (pH = 7.4), 4–5 hours at 39,000 rpm. Spinco, model L. ● − optical density at 260 nm: O − radioactivity.

Koshcheenko and his colleagues [65] also examined the effect of overall gamma irradiation at the minimum lethal dose on the biosynthesis of different types of cell RNA in the liver. They found that the specific activity of mRNA, rRNA and cytoplasmic RNA rises markedly during the early stages of radiation sickness. Other workers reported that irradiation at a dose of 50–900 r blocks the biosynthesis of nuclear RNA in the liver [66].

POSSIBLE CAUSES OF THE EFFECT OF IRRADIATION ON RNA

Results obtained with regard to the effects of irradiation on RNA are highly contradictory and have therefore been interpreted in various ways. According to Libinzon [67], inhibition of RNA synthesis cannot be attributed to a deficiency of phosphorylated precursors. Moreover, no changes could be detected in the activity of RNA polymerase in the irradiated tissues [68]. The activity of RNA polymerase in rat liver following X-ray irradiation of the animals at a dose of 750 and 1,500 r was found to be unchanged at different time intervals after a partial hepatectomy [69].*

* Data on activation of RNA polymerase in the liver of irradiated mice were published recently [Gatvin, M.B., Experientia, 26:490. 1970; Omata, S. et al., J.Biochem., 63:695. 1968]. However, irradiation did not increase the activity of RNA polymerase in adrenalectomized mice.− The Editor.

On the other hand, many workers reported that relatively low irradiation doses [70—74] markedly increase the template activity of DNA in the DNA- and RNA-polymerase systems [70—74]. According to Kuzin and Budilova [71] irradiation markedly suppresses activation of thymidylate kinase, induced by partial hepatectomy. In a study of the effect of gamma irradiation on the phosphorylation of thymidine in two radiosensitive tissues — rat thymus and spleen — Fillipovich [75] found that this process is markedly depressed (by 40%) immediately after irradiation, possibly owing to inhibition of the synthesis of thymidylate kinase.

Damage to the DNA template exerts a marked effect on RNA synthesis. The reduced template activity indicates that irradiation inactivates the template, apparently through a variety of structural lesions caused by the radiochemical modification of the template [41, 75—77]. Radiochemical changes in the DNA may lead to the formation of abnormal template RNA [25—78].*

As noted above, the synthesis of RNA continues almost at the same rate as before despite the numerous structural lesions and the fact that DNA synthesis is reduced by half. The synthesis of RNA was generally regarded as radioresistant in comparison with the synthesis of DNA. This view was based on studies showing that irradiation does not affect the amount of ribosomal RNA, which forms the bulk of the cell RNA. The course of research into the effect of radiation on RNA altered with the recent discovery of the existence of several types of this compound and more heterogeneous data were obtained. Briefly, the results depended on the type of cell RNA examined. Some workers found that RNA synthesis is unaffected [33, 34, 48] or even enhanced [46, 51—54] by irradiation, while others [47, 66, 81, 82] observed that irradiation exerts an immediate inhibitory effect on the synthesis of DNA-like RNA. These results indicate that radiation disturbs control of the types of RNA being synthesized.

Together with a group of colleagues we began in 1963 a study of the effect of radiation on RNA synthesis in seed embryos. The work was done with pea seeds, the cells of which proved to be synchronized at the G_1 period during the first 24—28 hours of swelling and represented a differentiating system with general mechanisms for the genetic control of RNA synthesis. We analyzed the population of newly formed types of RNA by MAK chromatography and ultracentrifugation in sucrose and cesium chloride gradients combined with hybridization techniques and determination of the nucleotide composition of the hybrid areas. Analysis of the effect of irradiation on the synthesis of different types of high molecular weight RNA in pea embryos revealed marked differences between them. The synthesis of the A—U type of RNA, apparently mRNA, was inhibited, while the breakdown of the high molecular weight precursor of ribosomal RNA to ribosomal RNA was accelerated and the synthesis of ribosomal RNA enhanced [79, 80]. The rate of synthesis of the A—U type of RNA is much lower (by 36—40%) in the embryos of the irradiated seeds than in control seeds at the

---

* In a recent work Hagen and his colleagues (Hagen, U. et al., Biochem. et biophys. acta 199 (1970), 115 showed that irradiated DNA binds RNA polymerase 1.6 times more than nonirradiated DNA. These workers suggest that the RNA polymerase is bound at single breaks in the DNA, thus blocking the synthesis of RNA along the DNA chain.— The Editor.

corresponding stage; this finding agrees with the data obtained with respect to the effect of radiation on RNA synthesis in other objects [81, 82].

Thus, the changes observed in the synthesis of newly formed RNA fractions indicate a radiation-induced disturbance of control at the level of transcription.

THE EFFECT OF RADIATION ON GENE ACTIVATION AND THE CONTROL OF GENETIC ACTIVITY

Study of the effect of radiation on the control of genetic activity is only just beginning. Most of the available data concerning the effect of radiation on gene activation relate to the synthesis of adaptive enzymes. Pollard and his colleagues [83, 84] found that irradiation suppresses induction in E . c o l i cells. In a study of formation of the adaptive enzyme β-galactosidase in irradiated bacterial cells, they observed a slight initial increase in the amount of this enzyme, followed by a distinct reduction in its synthesis. Comparative study of the effect of gamma radiation, deutrons and alpha particles on the synthesis of this protein revealed that the process is markedly radioresistant. Other authors [85] found no activation of β-galactosidase in a cell-free system isolated from irradiated cells of E . c o l i. X-ray irradiation apparently blocks the synthesis of mRNA by inactivating the DNA template, since inhibition can be eliminated by adding DNA from induced cells. These authors point out that the induced synthesis of β-galactosidase is more sensitive to ionizing radiations.

A radiation-induced inhibition of the induction of bacterial enzymes is also reported in other publications [86–88], which stress that the effect obtained depends on the radiosensitivity of the strain and composition of the growth medium. However, no correlation was observed between suppression of enzyme induction and the survival rate. Suppression of synthesis of the induced enzyme is invariably greater than that of protein synthesis in general. The high radiosensitivity of the synthesis of inducible enzymes was also demonstrated in animals [89].

The synthesis of some enzymes in the liver of newborn rats is inhibited by in utero irradiation. Experiments conducted in Du Bois' laboratory [90, 91] showed that X-ray irradiation inhibits almost completely the activation of several enzyme systems, normally taking place in male rats aged 30–50 days under the influence of androgens. Inhibition of enzyme synthesis was also obtained by irradiating the head, which suggests that the radiation effect involves the hormonal system.

Interesting results were obtained in a study of the effect of irradiation on the synthesis of RNA in regenerating liver [92, 93]. Concerning the biochemistry of the interphase in regenerating liver it is known that there is early stimulation of the synthesis of RNA and protein at the beginning of the cycle (G_1 period); in other words, partial hepatectomy activates a large group of genes during the first hours after the operation. These events are followed by the appearance of thymidylate kinase and DNA polymerase, but DNA synthesis does not begin until 15–18 hours after hepatectomy.

The appearance of enzymes engaged in DNA synthesis is accompanied by a drop in activity of the enzymes catalyzing the breakdown of pyrimidines [92]. It follows from these findings that irradiation inhibits activation of RNA synthesis in regenerating liver cells during the first 6 hours after the operation but does not affect the rate of RNA synthesis 18 hours after it [93]. The effect of radiation on the different stages of regeneration may thus be outlined as follows. If irradiation takes place before the appearance of the enzymes necessary for DNA synthesis, it delays the formation of these enzymes, on the other hand, if irradiation is carried out at a time when the enzymes are already present, the effect on DNA synthesis is negligible. Comparison of these findings with the stimulatory effect of irradiation on RNA synthesis in liver cells leads to the conclusion that irradiation inhibits activation of new genes.

A correlation between the effects of irradiation and actinomycin D has been observed in various enzyme systems. Thus, activation of thymidylate kinase by partial hepatectomy or unilateral nephrectomy is prevented absolutely by exposure to actinomycin D or X-ray irradiation at a dose of 1,000 r [94—96]. Mishkin and Shor [96] found that actinomycin inhibits induction of tryptophan pyrrolase by cortisone. This effect appears to be unrelated to radiation-induced damage to a template. According to Kuzin and Mel'nikova [97], irradiation inhibits induction of tryptophan pyrrolase by tryptophan, whereas addition of cortisone causes synthesis of the normal enzyme.

Thus, irradiation inhibits induction whenever activation of the enzyme by a specific inducer is associated with the synthesis of mRNA (except when enzyme activation is due to hormones) [96—98]. However, the data available on the various mechanisms of gene activation cannot yet explain the differing radiosensitivity of induction by substances and hormones.

Another aspect of the effect of radiation on the control of genetic activity concerns radiosensitivity of different periods of the cell cycle. It is known that transition of the cell from one period to another involves activation of genes and synthesis of specific forms of RNA. Many workers have demonstrated the high radiosensitivity of the transition phase from one stage of the cell cycle to the next, expressed, for example, in inhibition of the passage of the cell from the G_1 to the S period [40, 99] and delay in mitosis [100—102]. Interesting data on the high radiosensitivity of changes in the information transfer from DNA molecules were obtained in Kuzin's laboratory [103].

Fradkin [104] stressed the greater radiosensitivity of the control mechanism in comparison with the programming one. He worked with the temperate phage λ, the genome of which contains a regulator gene and an operator gene ("control mechanism") in addition to its complement of structural genes. For this reason the temperate phage-lysogenic cell system proved a convenient biological model for studies of the comparative functional vulnerability of different parts of the genome without subsequent genetic analysis.

The modern conception is that gene activity controls the entire differentiation process. Attempts to determine the effect of radiation on mRNA and other elements of protein synthesis during the development and differentiation of higher organisms are therefore of particular interest. The

embryos of various amphibians and fish at different stages of development
were found to be convenient objects for such studies since only mRNA is
synthesized in them during this period [105−107].   Using radiation to inactivate the nuclei, Neifakh [108] examined the function of the nucleus during
the early development of fish.   He found that the relationship between the
time of irradiation and the time of developmental arrest varies from one
period to another.   Thus, irradiation at the early blastula stage (from 0 to
6 hours) stops embryonic development at the 9th hour of life; irradiation
at the middle gastrula stage stops the development of the late gastrula
stage, and still later the moment of growth arrest similarly depends on the
time of irradiation.   The periods between 6 and 8 hours and after 14 hours
were named periods of "morphogenetic activity of the nuclei." The author
assumed that the nucleus actively produces information and releases it to
the cytoplasm during these periods but is inactive at 0−6 and 9−14 hours.

Subsequently, Neifakh attempted to associate the periods of morphogenetic activity with the synthesis of mRNA [109].   Belitsina, Neifakh and
others examined the effect of irradiation on the synthesis of mRNA in loach
embryos.   Having found no changes in the synthesis of RNA after irradiation,
they concluded that the morphogenetic activity of the nucleus is not responsible for the differing radiosensitivity of the growth periods [110].   Some
time later Kafiani and his colleagues [81] found that irradiation of the
embryos inhibits by 30−60% the synthesis of mRNA at all stages of development.   This finding, however, did not explain the existence of periods characterized by differing radiosensitivity.   It was assumed therefore that
periods of "morphogenetic activity" may be associated with activation
of new genes.

Direct proof of the latter assumption was provided by the experiments of
Umanskii [111] in our laboratory, using the same material.   By means of
the competitive hybridization method, Umanskii examined the RNA spectrum
during early embryogeny of the loach both in normal organisms and after
gamma irradiation of the eggs at a dose of 30 hr.   He found a periodic activation of new genes during the early embryogeny of the loach.   Irradiation
during the period of activation of new genes does not cause appreciable repression of genes which are already functional.   On the other hand, irradiation inhibits activation almost completely (by 80%) if applied before the
beginning of this process.

It is known that actinomycin D is bound at certain sites on the DNA molecule and lowers the de novo synthesis of RNA.   The correlation between the
effects of radiation and actinomycin D has already been mentioned [94−98].
These data indicate that both gene induction and inhibition of this process
take place at the transcription level.   In other words, the mechanism by
which radiation disturbs regulation of the synthesis of RNA templates can
only be established by studying RNA at the time of its synthesis on DNA,
i. e., by analyzing the natural DNA−RNA hybrid complex after irradiation.

We tried to determine the effect of gamma radiation on the earliest
stages of RNA synthesis by isolating and analyzing the natural DNA−RNA
hybrid complex.   This approach has considerable advantages over the
analysis of free RNA of cell origin since it shows the rate of synthesis of
the different types of RNA irrespective of the rate of their breakdown in

the cell, which may change markedly following irradiation and lead to erroneous results with regard to the synthesis proper [48]. In the absence of DNA replication (G_1 period), two molecular forms of DNA were found in plant embryo cells — native and partly denatured. By isolating the partly denatured fraction it was possible to determine the physical state, nucleotide composition and metabolic activity of these molecules [112, 113]. The isolated natural hybrid was even divided into fractions of different nucleotide composition, i. e., RNA of the G—C and A—U types [114, 115]. In a sucrose gradient RNA associated with DNA forms several fractions with different sedimentation coefficients. In a cesium chloride gradient the nascent RNA bands together with DNA; the hybrid molecules have a greater density than those on which RNA synthesis does not take place.

Studies of the effect of radiation on the natural DNA—RNA hybrid complex (Figure 4.5) revealed that the hybrid isolated from irradiated embryos is less radioactive and less heterogeneous in the zone corresponding to DNA fractions bearing RNA of the A—U type, in comparison with the hybrid isolated from the control embryos [117, 118]. It appears, therefore, that the adverse effect of irradiation on the de novo formation of A—U type RNA [79—82] results not only from inhibition of synthesis but also from involvement of a smaller part of the DNA genome in the irradiated embryos.

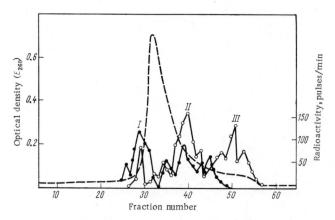

FIGURE 4.5. Comparison of chromatographic profiles of the natural DNA-RNA hybrid in normal seeds (○) and seeds exposed to gamma irradiation at a dose of 20 kr (●). The seeds were swollen in water for 16 hours after irradiation and then the embryos were incubated for 3 hours in a solution of uridine-C^{14}:

I—III) different types of newly formed RNA; — — — — optical density at 260 nm.

This assumption was unequivocally confirmed in comparative studies of the RNA population of normal and irradiated embryos by artificial hybridization. It was found that the drop in the de novo formation of mRNA, revealed by MAK chromatography, is associated with the fact that the irradiated embryos lack 40% of the RNA present in normal embryos at the corresponding stages of development. Moreover, the specific activity of the mRNA synthesized at the moment of irradiation is lower by 15—20% [119]. Further research showed that irradiation largely or completely represses activation of new genes [111]. There is practically no repression of already functional genes and the complement of RNA molecules found in the irradiated embryos is the same as in normal embryos at a stage corresponding to that of irradiated embryos at the moment of irradiation. In irradiated pea seeds, for example, part of the normally functional DNA genome is inactive during the first 20—28 hours of swelling, but at the same time incorporation of labeled precursor takes place enhancing the synthesis of ribosomal RNA.

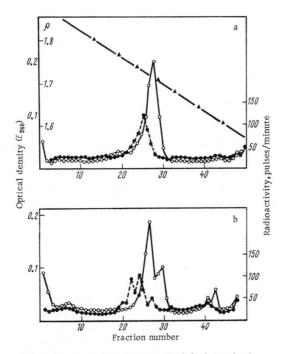

FIGURE 4.6. Analysis of the DNA-RNA hybrid complex in nonirradiated (a) and irradiated (b) embryonic cells in a cesium chloride density gradient. Spinco, model L, rotor SW-39, 37,000 rpm, 50 hours at 20°C. Density of the basal solution 1.704:

●) radioactivity; O) optical density; ▲) density of cesium chloride (ρ).

The assumption of a radiation-induced derepression of the template and the possible involvement of previously nonfunctional cistrons was confirmed in a study of the DNA—RNA complex by gradient ultracentrifugation [120, 123]. It became clear that irradiation increases the number and magnitude of the radioactivity peaks of a hybrid which in terms of nucleotide composition and sedimentation properties corresponds to ribosomal RNA (Figure 4.6). Indirect evidence that radiation causes derepression of ribosomal cistrons was provided by the cytological observations of V. V. Khvostova [121], who detected marked enlargement of the nucleolus in irradiated cells of the meristem of wheat and pea roots.

Earlier we made a comparative study of the growth and development of seedlings following exposure of the seeds to radiation of differing ionization density [8]. The early stages of development of embryonic shoots of seeds irradiated with fast neutrons and gamma radiation at more or less equal doses (taking into account the biological efficiency of neutrons, which is approximately 10) proceed uniformly during the first 3 days, apparently because of the already existing mRNA. Later on the embryos treated with gamma quanta develop to some extent and become green, although they lag far behind the controls in growth. Embryos in the neutron-irradiated seeds grow to a length of 2—3 cm, but then the majority die. By the 15th day all the neutron-treated seedlings have died, whereas those developing from gamma-irradiated seeds show a tendency to renew the growth of side rootlets These studies revealed an interesting fact: incorporation of $C^{14}O_2$ in the seedlings after irradiation of the seeds with an absolutely lethal dose (neutrons, 10 krad) completely inhibiting synthesis of DNA. We regard this phenomenon as a consequence of radiation-induced disturbance of the regulation of RNA synthesis. Since some of the information-bearing templates are lacking in the irradiated seedlings, their metabolic partner — ribosomal RNA synthesized on additional derepressed genes — is apparently unable to ensure the synthesis of new proteins and the seedlings die.

CONCLUSION

The processes involved in the transfer of new information occupy a central place in radiobiological research. Analysis of the main radiobiological effects on the informational function of DNA (differentiation processes and the processes of induction by substrates or hormones) shows that radiation causes specific disturbances in the functioning of genes.

Irradiation blocks transition from one stage of the cell cycle to the next in the course of development; it inhibits the synthesis of inducible enzymes much more than protein synthesis generally; it suppresses activation of thymidylate kinase (caused by partial hepatectomy); it inhibits appearance of ecdysone, the molting hormone of insects, at a specific developmental stage [122]; hormonal induction, however, is a radioresistant process; irradiation also prevents activation of some genes during embryogeny and disturbs regulation at the transcription level, preventing activation of some genes and causing derepression of others.

Thus, the processes associated with switching on and off of genes accompanying changes in the transmitted information have been found to be extremely radiosensitive. At the same time, experiments with a variety of objects show that there is apparently no pronounced repression of already functional genes, although RNA synthesis is somewhat inhibited quantitatively in both cases. These findings demonstrate radioresistance of the genetic components directly involved in the information transfer (that is, relative radioresistance of the structural genes in comparison with the regulator genes). The adverse effect of radiation on transcription control appears to be a complex phenomenon comprising several mechanisms.

BIBLIOGRAPHY

1. Euler,V.H. and G.Hevesy.—K.Danske Videnk. Selsk.,Biol. Med. 17 (1942),1.
2. Hevesy,G.— Advances Nucl. Acid Related Subjects Biochem. 7 (1948),111.
3. Stocken,L.A.— Radiat. Res. Cited from A.M.Kuzin. "Radiatsionnaya Biokhimiya," p.192. Moscow, AN SSSR,1962.
4. Lajthal,L.G. et al.— Radiat. Res. 8 (1958),1.
5. Holmes,B.— Sympos. on Ioniz. Radiat. on Cells.,p.131. L.—N.Y., Academic Press,1960.
6. Vainson,A.A. and A.M.Kuzin.— Doklady AN SSSR 165 (1965),933.
7. Kuzin,A.M. and A.A.Vainson.— Radiobiologiya 5 (1965),785.
8. Tokarskaya,V.I. and A.M.Kuzin.— Radiobiologiya 6 (1966),1.
9. Howard,A. and S.R.Pelc.— Heredity,Lond. 6 (1952),261.
10. Holmes,B.E. and L.K.Mee.— Radiol. Symp. Ed. Bacq, Z. and P.Alexander,p.220. London, Academic Press,1959.
11. Kelly,L.S.— Proc. Second United Nat. Int. Conf. Peaceful Uses Atom. Energy,U.S.A.,p.886. 1961.
12. Terasima,T. and Z.L.Tolmach.— Science 140 (1963),490.
13. Khollender,A.V.— In: "Pervichnye i nachal'nye protsessy biologicheskogo deistviya radiatsii," p.150. Moscow, AN SSSR,1963.
14. Errera,M.— In:"Radiobioloigy," p.151. Moscow, I.L.,1955. (Russian translation)
15. Bollum,F.S. et al.— Cancer Res. 20 (1960),138.
16. Creasey,W.A. and L.A.Stocken.— Biochem. J. 69 (1959),17.
17. Romantsev,E.F. et al. Early Radiation-Biochemical Reactions.— Moscow, Atomizdat,1966.
18. Libinzon,R.E.— Proceedings of the All-Union Conf. on Radiological Medicine. Eksp. med. radiol. Moscow, Medgiz,1957. (Russian)
19. Nygaard,O.F. The Effects of Ioniz. Radiations on Immune Processes. Ed. Charles, A.Leone, Gordon and Breach,p.47.— N.Y. Sci. Publishers,1962.
20. Chambon,P. et al.— Life Sci. 5 (1962),167.
21. Berg,J. and M.Goutier.— Arch. Int. Phys. Bioch. 75 (1967),36.
22. Kuzin,A.M. The Structural-Metabolic Hypothesis in Radiobiology.— Moscow, "Nauka." 1970. (Russian)
23. Strazhevskaya,N.B. and V.A.Struchkov.— Radiobiologiya 2 (1962),9.
24. Struchkov,V.A.— In: "Nukleinovye kisloty i biologicheskoe deistvie ioniziruyushchei radiatsii," p.56. Moscow, "Nauka," 1967.
25. Spitkovskii,D.M. et al.— Ibid.,p.7.
26. Scholes,G.— Prog. Biophys. Molec. Biol.,Vol.13,p.59. Oxford—London—New York—Paris Press,1963.
27. Shal'nov,M.I.— Thesis. Pushchino-na-Oke,1968.
28. Tokarskaya,V.I. and P.A.Nelipovich.— Radiobiologiya 7 (1967),938.
29. Khalikov,D.R. et al.— Second All-Union Biochemical Congress. 17th Section-Radiation Biochemistry, p.24. Tashkent,1969. (Russian)
30. Tokarskaya,V.I. The Initial Stages of Radiation Injury to DNA.— Second Intern. Congr. Radiat. Res. Abstr.,p.15. Harrogate, Yorkshire,England,1962.
31. Tokarskaya,V.I.— Radiobiologiya 1 (1961),2.
32. Holmes,B.E.— Br. J. Radiol. 20 (1947),450.
33. Mil'man,L.S. and N.M.Shapiro.— Radiobiologiya 2 (1962),530.
34. Belitsina,N.V. et al.— Doklady AN SSSR 153 (1963),1204.
35. Kafiani,K.A. et al.— Biokhimiya 31 (1966),365.
36. Mandel,P. et al.— Proc. 2nd Int. Conf. Peaceful Uses Atom. Energy, Geneva,1958.
37. Mandel,P. and P.Chambon. Immediate and Low Level Effects of Ionizing Radiation,p.71.— London, Taylor and Francis,1960.

38. Dikovenko,E.A.— Proc. All-Union Conf. on Radiological Medicine and Experimental Medicine,p.96. Moscow,Medgiz,1957. (Russian)
39. Kalacheva,V.Ya.— Thesis. Moscow,1957.
40. Pelc,S.R. and A.Howard.— Radiat. Res. 3 (1955),135.
41. Kuzin,A.M. and V.I.Tokarskaya.— Biofizika 4 (1959),446.
42. Budnitskaya,E.V. et al.— Radiobiologiya 7 (1967),133.
43. Payne,A.H. et al.— Proc. Soc. Exptl. Biol. and Med. 81 (1952),698.
44. Smellie,R.M. et al.— Biochem. J. 60 (1955),177.
45. Harbers,E. and C.Heidelberger.— J. Biol. Chem. 234 (1959),1249.
46. Mant'eva,V.L. et al.— Voprosy Meditsinskoi Khimii 12, No.4 (1966),407.
47. Budnitskaya,E.V. et al.— Voprosy Meditsiny 10, No.2 (1964),179.
48. Kalendo,G.S.— Thesis. Moscow,1966.
49. Cardella,J.M. and E.J.Lichtler.— Cancer Res. 15 (1955),529.
50. Kelly,L.S. et al.— Radiat. Res. 2 (1955),490.
51. Cejka,L. and J.Nosek. — Čas. českých Lék. 6 (1958),205.
52. Gubin,G.D.— Radiobiologiya 2 (1962),553.
53. Cejka,L. and J.Nosek.— Meditsinskaya Radiologiya 4, No.12 (1959),21.
54. Dikovenko,E.A.— Proc. All-Union Conf. on Radiological Medicine,p.96. Moscow,Medgiz,1957. (Russian)
55. Sharobaiko,V.I.— Tsitologiya 6 (1964),101.
56. Dikovenko,E.A.— Meditsinskaya Radiologiya 3, No.6 (1958),51.
57. Yamamoto,Y.L. et al.— Radiat. Res. 21 (1964),36.
58. Petrovic,S.— Biochim. biophys. Acta 61 (1962),842.
59. Petrovic,S. et al.— Bull. Scient. Cons. Acads. RPF Yugosl. 6 (1965),185.
60. Petrovic,S. et al.— Proceedings of the 5th Yugoslav. Conf. Radiobiology,Lublana,June,1964,p.57. Belgrade,Published Yugoslav. Nuclear Energy Commission,1964.
61. Hudnik Plevnic,T.A.— Biochim. biophys. Acta 103 (1965),515.
62. Simic,M. et al.— Third Int. Congr. of Radiat. Res.,Book Abstracts,p.203. Italy,Cortina d'Ampezzo,1966.
63. Ugarova,T.Yu.and P.H.Tseitlin.— Byull. Eksp. Biol. Med. 7 (1964),55.
64. Paskevich,N.F. et al.— Voprosy Eksper. Klinich. Radiol.,Vol.5,p.24. Kiev, "Zdorov'e," 1969.
65. Koshcheenko,N.N. et al.— In: "Nukleinovye kisloty i biologicheskoe deistvie ioniziruyushchikh izluchenii," p.186. Moscow, "Nauka," 1967.
66. Logan,R. et al.— Biochim. biophys. Acta 32 (1959),147.
67. Libinzon,R.E.— Inform. Byull. Radiobiologiya, No.6 (1964),53.
68. Chambon,P. et al.— Biochim. biophys. Acta 157, No.3 (1968),504.
69. Harrington,H.— Proc. Natn. Acad. Sci. U.S.A. 51 (1964),59.
70. Zimmerman,F. et al.— Biochim. biophys. Acta 87 (1964),160.
71. Budilova,E.V. and A.M.Kuzin.— Synopses of Reports of the Conf. on Mechanisms of the Biological Effect of Ionizing Radiation,p.86. Lvov,L'vovskii Inst.,1965.
72. Zimmerman,F. et al.— Biochem. Z. 342 (1965),115.
73. Budilova,E.V.— Radiobiologiya 7 (1967),323.
74. Weis,J.J. and C.M.Wheeler.— Biochim. biophys. Acta 145 (1967),68.
75. Fillipovich,N.V.— In: "Nukleinovye kisloty i biologicheskoe deistvie ioniziruyushchikh izluchenii," p.119. Moscow, "Nauka," 1967.
76. Okada,S.— Nature 185 (1960),193.
77. Wheeler,C.M. and S.Okada.— Intern. J. Radiat. Biol. 3 (1961),25.
78. Hagen,U. et al.— Third Intern. Congr. Radiat. Res, Book Abstr.,p.101. Italy,Cortina d'Ampezzo,1966.
79. Tokarskaya,V.I. et al.— Doklady AN SSSR 176 (1967),211.
80. Tokarskaya,V.I. and S.R.Umanskii.— Radiobiologiya 8 (1968),1.
81. Kafiani,K.A. et al.— Biokhimiya 31 (1966),365.
82. Trams,G. and I.Vollertsen.— Third Intern. Congr. of Radiat. Res.,Book Abstr.,p.223. Italy,Cortina d'Ampezzo,1966.
83. Pollard,E.C.— Intern. Symp. Inition. Effect on Ioniz. Radiat., Abstr.,p.59. Moscow,1960.
84. Pollard,E.C.— Science 146 (1964),927.
85. Novelli,G. et al.— J. Cell. Comp. Physiol.,Suppl. 1 (1961),225.

86. P a u l y , H.— Nature 184 (1959), 1570.

87. M o o r e , I.L.— J. Gen. Microbiol. 41 (1965), 119.

88. G i n o z a , W.— Ann. Rev. Nucl. Sci. 17 (1967), 469.

89. S m i t h , C.H. and M.L.S h o r e.— Radiat. Res. 29 (1966), 499.

90. H i e t b r i n k , B.E. et al.— Radiat. Res. 16, No.4 (1962), 555.

91. Du B o i s , K.P.— Radiat. Res. 30 (1967), 342.

92. B o l l i m , J. and I.A n d e r e g g.— Cancer Res. 20 (1960), 138.

93. W e l l i n g , W. and J.A.C o h e n.— Biochim. biophys. Acta 42 (1960), 181.

94. F a u s t o , N. et al.— Archs. Biochem. Biophys. 106 (1964), 447.

95. B e l t z , R.E. et al.— Biochem. Biophys. Res. Commun. 1 (1959), 298.

96. M i s h k i n , E.P. and M.Z.S h o r e.— Biochim. biophys. Acta 138 (1967), 169.

97. M e l ' n i k o v a , S.K. and A.M.K u z i n.— Radiobiologiya 9 (1969), 2.

98. K u z i n , A.M. et al.— Radiobiologiya 7 (1967), 3.

99. S h e r m a n , G. and H.Q u a s t l e r.— Expl. Cell Res. 19 (1960), 343.

100. P a i n t e r , R.B. and J.S.R o b e r t s o n.— Radiat. Res. 11 (1959), 206.

101. K i m , J. and T.C.E v a n s.— Radiat. Res. 21 (1964), 129.

102. F r a n k f u r t , O.S. and L.P.L i p c h i n a.— Doklady AN SSSR 154 (1964), 207.

103. K u z i n , A.M.— In: "Nukleinovye kisloty i biologicheskoe deistvie ioniziruyushchikh izluchenii," p.150.
     Moscow, "Nauka." 1967.

104. F r a d k i n , G.E.— Ibid., p.14.

105. S p i r i n , A.S. et al.— Obshchaya Biologiya 25 (1966), 321.

106. G l i š i n , V. and M.G l i š i n.— Proc. Natn. Acad. Sci. U.S.A. 52 (1964), 1548.

107. K a f i a n i , K.A. and M.Ya.T i m o f e e v a.— Doklady AN SSSR 154 (1964), 721.

108. N e i f a k h , A.A.— Zh. Obshch. Biol. 20 (1959), 202.

109. N e i f a k h , A.A.— Zh. Obshch. Biol. 22 (1961), 42.

110. B e l i t s i n a , N.V. et al.— Doklady AN SSSR 153 (1963), 1204.

111. U m a n s k i i , S.R.— Radiobiologiya 7 (1968), 2.

112. T o k a r s k a y a , V.I. et al.— Biokhimiya 33 (1968), 542.

113. T o k a r s k a y a , V.I. and S.R.U m a n s k i i.— Molekulyarnaya Biologiya 1 (1967), 511.

114. T o k a r s k a y a , V.I. et al.— Molekulyarnaya Biologiya 2 (1968), 420.

115. T o k a r s k a y a , V.I.— Doklady AN SSSR 175 (1967), 733.

116. T o k a r s k a y a , V.I. et al.— Molekulyarnaya Biologiya 3 (1969), 527.

117. T o k a r s k a y a , V.I.— Radiobiologiya 7 (1967), 480.

118. T o k a r s k a y a , V.I.— Doklady AN SSSR 180 (1968), 482.

119. U m a n s k i i , S.R. and V.I.T o k a r s k a y a.— Radiobiologiya 8 (1968), 2.

120. T o k a r s k a y a , V.I.— Doklady AN SSSR 191 (1970), 464.

121. K h v o s t o v a , V.V. et al.— Tsitologiya 3 (1961), 183.

122. K u z i n , A.M. and I.I.Y u s i f o v .— Radiobiologiya 7 (1967), 3.

123. T o k a r s k a y a , V.I.— Studia biophysica 15/16 (1969), 63.

Chapter 5

THE STRUCTURAL AND FUNCTIONAL ORGANIZATION
OF THE CHROMOSOME AND THE ROLE OF RADIATION-
INDUCED LESIONS OF ITS DNA

THE STRUCTURAL AND FUNCTIONAL ORGANIZATION
OF THE CHROMOSOME OF MULTICELLULAR
ORGANISMS

Life is based on the specific chemical catalytic activity of the fundamental cell structures known as chromosomes, which consist essentially of
DNA — a compound capable of autoreproduction by convariant replication
[1–3]. Available radiobiological evidence indicates that the cell chromosome is one of the main targets of the lethal and genetic effects of radiation
in living organisms [4–5].

There is a great difference between DNA in aqueous solution as a substrate of radiolytic reactions and DNA in chromosomes as a substrate of
lethal or premutational lesions of the living cell or organism. For this
reason any discussion of the effects arising from lesions of nucleic acids
in the organism in terms of molecular radiobiology must begin with a description of the structural and functional organization of the chromosome,
and in particular that of multicellular organisms.

The exact structural and functional organization of the basic chromosomal material of multicellular organisms — DNA and protein — is still
unknown, although research carried out over many years has revealed a
number of fine details of the microscopic and submicroscopic structure of
chromosomes during the different periods of their mitotic and meiotic
cycles as well as of their structure in certain nondividing cells [6, 7].

Research into the finest structural and functional details of the chromosome is impeded by various known technical difficulties. The consecutive
stages of condensation of the chromatin in interphase cells remain outside
the resolving capacity of both electron and light microscopy; the electron
micrographs of interphase chromosomes as "naked" DNA or elementary
chromosomal threads of various thickness do not "fit" the light photomicrographs of metaphase and anaphase chromosomes. Hence the inclination of
many workers to use models in an attempt to combine the biochemical,
cytological and genetic data into an integral picture and present the chromosome as a functional entity, the structure and functions of which can be
studied at the molecular level [8].

The inevitable subjectivity in interpretation of the results has led to
diversity in the models. The same experimental data have been used to
elaborate various models which may be classified into three categories.

1) The chromosome represents a regular arrangement of DNA molecules or nucleoprotein particles joined together by units not containing DNA or attached alternately at the ends to a short central axis.

2) The axis of the chromonema consists of a multichain complex of DNA, protein and other components with numerous double helices of DNA.

3) The visible chromonema of a typical chromosome is composed of a single protein-bound DNA double helix assembled and coiled according to a specific program [9].

This chapter deals with current concepts of the state of chromosomal material, processes of synthesis, coiling and uncoiling of DNA, mechanisms of chromosomal reproduction, and transfer of genetic information taking into account recent evidence and existing models.

Cytological studies of spiderwort showed long ago that the chromatin located peripherally along the body of the metaphase chromosome in pollen mother cells has a helical form [10]. Similar structures, giving the chromosome the appearance of a sequence of alternating dark and light discs, were found in a variety of cytological objects including metaphase chromosomes of mosquito larvae. The hypothesis that the chromosome contains a spiral thread or chromonema found some support in the cytological analysis of anaphase somatic chromosomes of spiderwort: a light colored substance is visible between the chromatin threads, which are twisted round one another. In cross section the anaphase chromosomes appeared as a dark ring with a light center, while intact they took the form of a rod with a dark periphery and a light core [11]. The helical threads inside the chromosome were still visible during telophase and did not vanish from the field of vision before interkinesis. The bands of chromatin threads reappeared during prophase, and the cycle of coiling and uncoiling was resumed. Condensations termed chromomeres were found on the spiral chromatin threads. During early prophase the chromomeres are arranged in a single row almost on the surface of the elongated cylinder formed during metaphase, however, when the chromosomes become shorter and thicker, the chromomeres form a spiral pattern like a zigzag or helix [12].

Further cytological studies of meiotic and mitotic chromosomes of spiderwort by light microscopy revealed several levels of coiling: in meiosis, two clearly distinguishable spirals, one large, containing 10 to 30 turns, the other fine, situated perpendicularly to the large spiral and consisting of numerous small turns; in mitosis, a large spiral which develops progressively from prophase to metaphase and causes the chromosome to contract to $^1/_{12}$ its length during prophase [13]. In cultured pollen cells of the lily and also in cells treated with agents which loosen the spirals the chromosomes in the first metaphase of meiosis were found to contain a very fine spiral with turns 30−50 nm in diameter, as well as the large and small spirals. This extremely fine spiral is present in the chromonemata in the early prophase of meiosis and does not vanish during conjugation of the chromosomes, i. e., in the leptonema, contrary to the assumption that the chromosomes are totally uncoiled during this period. The concepts established by 1935 as regards the respective location of the chromatin and achromatin components of the chromosome were summed up by Heitz [8] in a scheme of the anaphase chromosome which is still essentially accepted today [7].

The nucleoproteid structure of the chromosome was revealed long ago largely through the successful application of ultraviolet spectroscopy [14, 15]. However, the function of the deoxyribonucleic and protein components of the chromosome remained obscure for a considerable time. Thus, N.K.Kol'tsov [12] regarded the chromosome as a protein core composed of two chromonemata (genonemata bearing the complete gene complement of the chromosome) surrounded by an adhesive and functionally inert nucleoprotein matrix. He disagreed with Demerec [16], who believed that all genes are no more than varieties or simply isomers of thyminucleic acid. A similar protocentric model of the chromosome [17] has been adopted in biophysical analysis of the physiological and genetic lesions of chromosomes, induced by ionizing radiations [5]. This controversy was settled with the discovery of the genetic role [18] and fine helical structure [19, 20] of DNA.

Electron microscopy has provided valuable information on the fine structure of the chromosome [8]. However, attempts to reconstruct the macrostructure of the chromosome on the basis of electron micrographs of its subunits led away from the developing concept of the multistage (progressive) coiling of DNA in the chromosome and gave rise to multistranded chromosomal models based on progressive pairing — a concept contrary to the experience of nearly a century of cytomorphological research. If the chromosome as seen under the light microscope is about 320 nm thick and the elementary chromosomal thread viewed in the electron microscope has a thickness of only 10 nm, the discrepancy between these data can be bridged by visualizing the chromosome as a multicore cable formed by the pairing of 32 DNA threads. Such is the structural pattern of the chromosome as composed of 32 DNA threads arranged "side by side" in the models proposed by Hans Ris [8] and Steffensen [21]; a pair of double DNA helices 2 nm thick form a cord about 4 nm thick, two cords joined together give an elementary chromosomal strand 10−12 nm thick, two such strands form a quarter chromatid measuring 20−25 nm, two quarter chromatids form a half chromatid of about 50−75 nm, two half chromatids a chromatid of about 100−150 nm, and two chromatids a chromosome of about 200−300 nm.

The models proposed by Ris and Steffensen became widely known [7, 22]. Although highly ingenious, the multistranded chromosomal models proposed subsequently [23, 24] are rather artificial and bizarre versions of those described above.

Research into the structural and functional organization of the chromosomes of multicellular organisms entered a more advanced stage with the isolation of viruses and determination of the nucleoprotein composition of bacterial viruses and phages. Many workers saw a parallel between the genes of the complex nucleoproteid chromosomes and the nucleoproteid viruses and phages [1, 5]. By the late 1950s and the early 1960s this idea had evolved into a hypothesis according to which the chromosome of multicellular organisms is formed as a result of the union of numerous circular DNA molecules [25, 26]. Indeed, the replicating chromosomes of phages and bacteria are usually composed of a Watson-Crick DNA double helix variable in length (molecular weight) and closed to form a single ring. The circular DNA of E. coli has a molecular weight of $2.5 \cdot 10^9$ daltons [27−29], and the circular DNA of the phage φX174 about $1.8 \cdot 10^6$ [30]. In some cases the bacterial chromosome may coexist with incorporated DNA of virus or phage origin. It is noteworthy that fragments isolated from irradiated chromosomes of multicellular organisms equivalent to one arm (produced by splitting of the centromere) or one replicon can reproduce in the cell in the same way as the entire chromosomes [7, 31]. True, in the latter case it has not been proved that there is no connection between the chromosome and the fragments. Possibly there is such a connection but invisible, since the acentric fragments do not usually participate in the poleward movement and disappear or at any rate do not remain as permanent satellites of the chromosome. The concept of the single-stranded, double-helical and multi-replicon structure of the chromosomes of multicellular organisms has been confirmed experimentally in a large number of studies [32–37].

Taylor [9] proposed an interesting model of the chromosome as a solitary double DNA spiral extending without interruptions throughout the entire length of the chromosome. This single-stranded model became well known in the USSR [38–40]. Different versions of the single-stranded model were proposed by Mosolov [41] and Polyakov [42].

The use of chromosomal models based on nearly a century of research in cytomorphology, electron microscopy, biochemistry, autoradiography, genetics and cytogenetics has provided a fairly objective general picture of the structural and functional organization of the chromosome of multicel-lular organisms. For a start, detailed observations of the structure of the chromosome by means of light microscopy have shown that packing of the chromatin is achieved by a multistage (progressive) coiling of DNA. The published photomicrographs show signs of plectonemic and paranemic coiling of the chromatin. Some data suggest the existence of a sinistrorse α-helix of DNA in the chromosomes, in addition to the dextrorse of α-helix. The possibility of a transition of the DNA α-helix from dextrorse to sinistrorse during replication greatly facilitates understanding of this mechanism, which is rather mysterious at the level of the complex chromo-some. Finally, although the DNA molecule is frequently broken into repli-cating subunits during replicon-by-replicon synthesis, this does not rule out the possibility of the life cycle of the cell including a stage in which the molecule of chromosomal DNA must be intact [33]. In mitosis, the G_2 period, prophase, metaphase, anaphase, and telophase are such stages. During these periods the double DNA helic constituting the basis of the chromonema presumably forms a ring, closed at the centromere to give a twisted loop. These concepts lead to a dynamic model reflecting changes in the DNA and protein of the chromosome during the major stages of the life cycle of the cell in mitosis and interphase.

The chromosome at the metaphase and anaphase of mitosis. Figure 5.1 shows schematically a metacentric (equal-armed) chromosome during the metaphase of mitosis [43]. One of the chromatids (on the left side) shows the overall arrangement of the coiled chromatin, while the other reveals the fine structure of the chromatin as it appears at all levels of the coiling of DNA. According to this model, each chromatid consists of two subunits named chromonemata (half chromatids), one chromonema having a dex-trorse DNA α-helix, the other a sinistrorse helix. Each chromonema consists essentially of two looped DNA macromolecules which meet at the centromere. The Watson-Crick double helix (secondary DNA structure) undergoes three further stages of coiling which may be described as follows: within the ring, it forms a spiral helix about 10 nm in diameter (tertiary structure) which is twisted into a plectonemic helix about 20−25 nm in diameter (quaternary structure); next, the appearance of a terminal fusion (telomere) between the dextrorse and sinistrorse chromonemata is followed by a very tight twisting of the loop, leading to chromonemal super-spirals (fifth order structures) which have a diameter of about 200−300 nm and are visible under the ordinary light microscope. In essence this is the system of progressive coiling. In the most detailed descriptions of chromosomes [6−13] it is pointed out that the metaphase chromosome of multi-cellular organisms consists as a rule of two chromatids which in turn are composed of half chromatids (chromonemata).

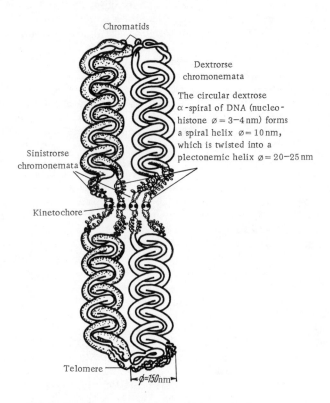

Chromatids

Dextrorse
chromonemata

The circular dextrose
α-spiral of DNA (nucleohistone $\varnothing = 3-4$ nm) forms
a spiral helix $\varnothing = 10$ nm,
which is twisted into a
plectonemic helix $\varnothing = 20-25$ nm

Sinistrorse
chromonemata

Kinetochore

Telomere

$\varnothing \approx 150$ nm

FIGURE 5.1. Diagram of a metacentric (equal-armed)
chromosome during the metaphase of mitosis. On the left −
general arrangement of the coiled chromatin in the
chromatid; on the right − the fine structure of the
chromatin: coiling of the looped DNA molecules, one a
sinistrorse α-helix, the other a dextrorse α-helix, constituting the dextrorse and sinistrorse chromonemata.

During the anaphase, chromatids of the metaphase chromosome containing isologous sets of genetic material diverge toward opposite poles after splitting of the centromeres to become the chromosomes of the two daughter cells. There is no sense in illustrating this stage since the anaphase chromosomes are in fact the separated sister chromatids of the metaphase chromosome and possess the chromatin structure shown in Figure 5.1. It only remains to mention that if the chromosome is defined as a structural unit of the cell containing one linkage group, this condition is also met by the chromatid of the metaphase chromosome (i. e., the anaphase chromosome), which therefore may be logically termed a chromosome. In anaphase chromosomes treated with acidic fixatives the slightly diverging chromonemata appear as two helices with intercalated whorls (9). Such a doubling is never observed in vivo; under natural conditions the anaphase chromatids take the form of solid cylinders. Such a segregation of paranemic helices could hardly be expected. This event would be most likely to occur when the intercalated chromonemata consist of two opposite helices − dextrorse and sinistrorse.

The controversy about the thickness of the elementary chromosomal thread as measured by electron microscopy is not surprising in view of the variety of methods used to obtain chromosome preparations. Gentle methods for the preparation of chromosomal material reveal helical fragments about 20−25 nm in diameter [8]; on the other hand, procedures such as microdissection, enzymatic microdigestion, treatment with potassium cyanide and with ultrasound yields finer spiral fragments about 10−12 nm in diameter, composed of threads 4 nm thick − apparently nucleohistone. Similar helical filaments were found also in the interphase nucleus of the ameba and in metaphase chromosomes of mammalian cells [9]. After being deproteinated in the interphase nucleus the filaments form so-called DNA plasma − a dehydrated form of naked DNA.

Thus, packing of DNA in the chromonema of the normal (nonpolytene) chromosome as shown diagrammatically in Figure 5.1 agrees with the findings of subcellular and macromolecular morphology; the diameter of the chromonema at the different levels of coiling is as follows on the basis of light and electron microscope measurements: nucleohistone 3−4 nm, spiral helix 10 nm, twisted spiral helix 20−25 nm, and chromonema 200−300 nm.

The 100- or 200-fold change in the length of the chromosome which takes place during the life cycle of the dividing cell according to cytomorphological observations may be logically explained in terms of progressive coiling and cyclization of chromosomal DNA. Cyclization of DNA can conceivably cause twofold shortening, the formation of a spiral helix a tenfold shortening, and a further tenfold shortening is obtained by formation of the visible superhelix of the chromonema. At the same time the thickness of the chromosome increases by a factor of 10−15 (from 10−20 to 100−300 nm). This satisfies the requirement for a constant density (constant volume) of the chromosomal substance according to the equation $L_1/L_2 = (D_2/D_1)^2$, where L_1 and L_2 are the lengths of the chromonema corresponding to the diameters D_1 and D_2. In other words, the shortened and thickened chromonema allows the packing of DNA by a linear (circular) rather than two-dimensional (side by side) or three-dimensional arrangement of the genes along the chromonema.

The circularly arranged genes become active only after uncoiling of the chromosome; supercoiling takes place only for a short period for the "technical purposes" of mitosis when the sister sets of genetic material (chromatids) have to be transferred to the daughter cells without losses.

With the finding by Kornberg [30] that the morphologically complete nucleoprotein corpuscle of the phage φX174 arises only after cyclization of its DNA, there is growing conviction that cyclization of DNA is a necessary condition for the coiling and supercoiling of all replicating nucleoprotein formations including such complex organelles as the chromosomes of multi- cellular organisms. The "normal" nucleoprotein package of the phage φX174 does not develop without enzymatic closure of the phage DNA into a ring. Morphogenesis of the nucleoprotein corpuscle of phage λ similarly requires cyclization of its DNA [44]: infection of E . c o l i cells with this phage is followed by initial increase in the amount of open circular DNA, which is the precursor of the closed supercoiled DNA. The latter appears to be neces- sary for formation of the nucleoprotein package of the phage particle. Phage DNA, coiled and organized into a tertiary structure which still lacks the head coat, can easily be seen under the light microscope as dense intra- cellular formations [45]. Cyclization and supercoiling precede the forma- tion of the protein coat around the DNA mass not only in the smallest phages but also in the largest ones [46, 47]. The single DNA molecule $(2.5 \cdot 10^9$ daltons) which constitutes the genome of E . c o l i is likewise closed into a ring [28, 29].

The circular structure of the chromosome of multicellular organisms is often masked, probably because the DNA ring which forms the basis of the chromonema is greatly extended in the form of a twisted loop. During 88 years of research into chromosomes, however, cytomorphologists have observed circular chromosomes also in multicellular organisms, even without attempting to study the cyclization process in particular. Thus, circular chromosomes were found in maize and some Diptera. A male line possessing a circular X-chromosome exists among Drosophila populations. This line was discovered by Lilian Morgan in the early 1930s. Soon after- ward a group of geneticists (Sokolov, Sidorov and Trofimov) proposed a novel hypothesis for the origin of the circular X-chromosome and managed to synthesize such a chromosomal complex according to a previously devised plan [12]. The circular chromosomes of plants and insects pass through a normal cycle of replication and cyclization even though they lack free ends [10, 13]. Bonner [48] referred cautiously to the possible exist- ence of circular chromosomes in the pea, stressing the difficulty associated with the packing of their DNA, the molecular weight of which is about $2.5 \cdot 10^{11}$ daltons or 100 times that of the chromosome of E . c o l i. The double helical structure of the DNA molecule was recently demonstrated by electron microscopy [49]. Attempts to detect cyclization of DNA were made during the electron microscopic isolation of salivary gland nuclei of mosquito larvae [50].

The circular structure of chromosomal DNA readily explains such mysterious parts of the chromosome as the centromere and the telomere. The centromere may be defined as the site of attachment of several chromo- nemal loops of different length (in the case of submetacentric and acrocentric chromosomes); in a way it is the mechanical center of the chromosome. The centromere contains DNA and is self-replicating, as is the chromosome as a whole. The centromere is also the point of attachment of the spindle fibers. The telomere, or terminal segment of the chromosome, can be visualized as the bunched ends of the looped chromonemata, held together by labile protein bonds. To prevent a terminal fusion of the chromosomes into rings the telomeres apparently require no special device other than the cyclization of DNA.

The chromosome during the G_1 period of interphase. Condensation of the chromatin continues until late anaphase; it is followed by loosening in the telophase and gradual unwinding of the coils which takes place simultaneously with formation of the nuclear membrane. During interphase the chromosomes uncoil to form threads invisible under the light microscope and fill up the newly formed nucleus. During this period the chromosome performs its biochemical functions in the cell as a regulator of protein synthesis. The structure of the functional chromosome at this time has been established by biochemical and autoradiographical research over the last 15 years; as a result of such research, the functions of the nucleic acid and protein components of the chromosome are now far better understood.

There is no doubt at present that DNA is both the structural and mechanical framework and the genetic determinant of the entire chromosome, contrary to the view existing until the late 1940s which interpreted its role as that of an adhesive, functionally inert matrix. The molecular weight of isolated and purified DNA from the most varied sources (10^6-10^{11} daltons) is many times greater than that of the heaviest polypeptides (about $2 \cdot 10^4$ daltons). Biochemical research is beginning to shed light on the morphological nature of the hetero- and euchromatin in normal interphase chromosomes and in giant polytene ones.

A morphological element of the functioning area of a normal interphase (G_1 period) chromosome is shown diagrammatically in Figure 5.2. This diagram combines the results of observations of the functioning of euchromatin puffs in polytene chromosomes of insects [6] and in lampbrush chromosomes, typically encountered during the growth period of ovocytes of birds, fish, reptiles and amphibians [6, 12, 51–53]. According to these observations the puffs of polytene chromosomes and the loops of lampbrush chromosomes result from uncoiling and deproteination of the chromonemal areas which synthesize RNA and protein, i. e., perform the functions of normally operating genes. The sites of the synthesis of RNA and protein around the lampbrush loop have been determined by the incorporation of tritium-labeled molecules of uridine and phenylalanine [6]: a complementary synthesis of RNA (transcription) takes place at the point of departure of the branches from the spiral helix, whereas polypeptides are synthesized on ribosomes near their point of re-entry. The invisible loops of DNA in polytene chromosomes of insect salivary glands function according to the same principle; the swellings in certain parts of these chromosomes correspond to uncoiling of the chromonemata and transition of their DNA to the state of template for the synthesis of mRNA. The puffs disappear with the completion of this process. According to current concepts [29, 48] the in vivo synthesis of mRNA involves only one of the DNA strands (the transcriptional strand) and necessitates disintegration of the other, inactive chain (in contrast to the corresponding process in vitro). The low weight of the DNA fraction noncomplementary to the mRNA can be explained by assuming that it includes fragments of the inactive strand, which has suffered numerous breaks at the sites of synthesis of mRNA, while the transcriptional strand remains intact during this process and is included in the heavy fraction. Protein synthesis is controlled according to the scheme of Jacob and Monod [54].

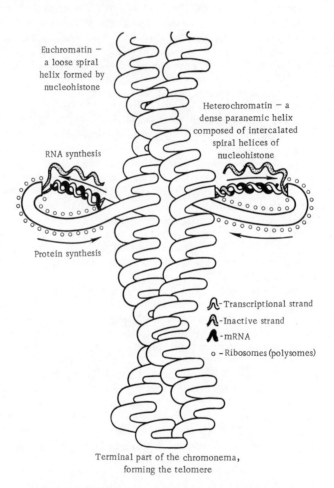

Euchromatin —
a loose spiral
helix formed by
nucleohistone

Heterochromatin — a
dense paranemic helix
composed of intercalated
spiral helices of
nucleohistone

RNA synthesis

Protein synthesis

Ⱂ-Transcriptional strand
Ⱂ-Inactive strand
Ⱂ-mRNA
o – Ribosomes (polysomes)

Terminal part of the chromonema,
forming the telomere

FIGURE 5.2. Diagram of a functional part of the DNA of a chromonemal loop during the G_1 period. Sites of complementary synthesis of RNA on DNA (transcription) are shown at the departure of the loops from the spiral helix. The synthesis of polypeptides (protein) takes place on polyribo-somes near the re-entry of the loops into the spiral helix.

The protein substance of the chromosomes of multicellular organisms consists of basic and acidic proteins as well as nuclear enzymes. The amount of basic proteins — histones — in the chromosome is roughly equal to that of DNA [55, 56]. There are two histone fractions — arginine-rich and lysine-rich, which presumably occupy different sites in the nucleo-histone fibril and perform different functions. It is thought that the arginine-rich histone, in a partly helical α-form, lies inside the large groove of the DNA double helix, whereas the lysine-rich histone occupies the small groove. Much remains to be discovered as to the manner in which the histone molecules are bound with one another and with molecular DNA, but some hypotheses based on autoradiographical, biochemical and electron microscopic data have been published [56]. The histone molecules may be bound with one another by mutual overlapping of the extended parts of the polypeptide chains at the ends of the protein molecules, where there are hydrogen bonds. The nature of the histone fraction varies from one section of DNA to another. The histones are evenly spread along the DNA chain in the native (repressed) nucleohistone (1:1 by mass); in contrast, the actively functioning parts of DNA contain only a slight amount of histone [57].

The histone links keep the spiral helix in a longitudinal position [58] and at the same time fasten the coils at the sites where these are in contact or inserted into one another, forming paranemic areas of heterochromatin along the plectonemic spiral loop. It appears on these grounds that the histone cross links occur at a frequency corresponding to one gene. From a cytomorphological and biochemical viewpoint the areas between these links can be regarded as replicons and may indeed act as such during the S stage. The heterochromatin microtubule 17 nm in diameter [58], formed by the regular coiling of the DNA, possibly corresponds to the paranemic spiral helix. The films of compact chromatin covering the membranes may consist of similar microtubules but in a less regularly packed state.

The role of histones in the maintenance of the chromatin structure was determined by electron microscopy of the nuclear chromatin involving stepwise extraction of histones from the nuclei of calf thymus lymphocytes [60]. It was found that removal of the arginine-rich histone (80% of the total histone) does not destroy the compact structure of the heterochromatin but extraction of the lysine-rich histone (20%) does. Addition of lysine-rich histone to the nuclei restores the dense structure of the hetero-chromatin (which is inactive as far as RNA synthesis is concerned). It is assumed that only lysine-rich histone forms cross links between the chromatin fibrils, whereas arginine-rich histone, though similarly bound to these fibrils, does not participate in such links.* The labile, readily broken bonds between arginine-rich histone and DNA are possibly of the salt type (that is, bonds between a polycation and a polyanion). The acidic proteins, which apparently form covalent bonds with DNA, are synthesized in the nucleus during the G_1 and G_2 periods [32]. Judging from the incorporation of tritium-labeled arginine [36], histone is synthesized twice as rapidly

---

* Studies of the structure and function of the DNP of the chromatin in animal cells are successfully underway (Molekulyarnaya Biologiya, 4:246, 291, 821. 1970). Experiments involving the stepwise extraction of DNP proteins from the chromatin have shown that the f_1 (lysine-rich) histone is responsible for inhibition of the activity of DNP as a template (Molekulyarnaya Biologiya, 5:586. 1971).— The Editor.

during the late S period and the G_2 period as in the G_1 and early S periods. According to De [32], the function of collagen-rich acidic proteins, like that of histones, is to maintain linear integrity of the chromosome. Nothing is known about the nature of the covalent bonds between the acidic proteins and DNA. The abundance of hydroxyamino acids in the so-called peptide fraction of DNA suggests that these bonds may be of the phosphoester or phosphoamide type [61].

Analysis of the factors controlling time of DNA replication and mRNA formation has led to the conception [62] that arginine-rich histone enhances the synthesis and replication of DNA as well as RNA synthesis, whereas lysine-rich histone inhibits these processes. It was also found [63] that histones inhibit the synthesis of DNA and RNA by forming complexes with DNA. These findings are not contradictory: the histone necessary for building the chromosome is synthesized at the same time as DNA replication is taking place or even before this, and without it the chromosome is not formed, whereas the histone participating in the complex with DNA affects activity of the latter as a template. Indeed, the synthesis of histones in polytene chromosomes of insects takes place at the sites of DNA synthesis and precedes the latter process [64]. Histone and DNA are synthesized during the same S period of the interphase in nuclei of regenerating liver; here the synthesis of nuclear RNA is accelerated immediately after DNA synthesis stops [65]. Judging from amino acid composition of the histones and nucleotide composition of the corresponding mRNA (A 31.8%; U 21.896%; C 23.1%; G 22.4%), synthesis of the latter can take place on only one transcriptional DNA chain [66], containing 31.8% thymine residues. It is noteworthy that histones and polylysine, which bind with the DNA double helix (as do actinomycin D, acridine orange and chloroquine), inhibit the synthesis of RNA, whereas the complementary RNA and RNA polymerase, which bind with single-stranded DNA (as do testosterone, estradial and methylcholanthrene) stimulate the synthesis of RNA. Arginine-rich histone apparently inhibits nucleic acid biosynthesis at certain concentrations only. Thus, the concentration of this compound in rabbit papilloma is no more than $\frac{1}{20}$ the value found in normal cells, and this reduction is regarded as responsible for the rapid synthesis of DNA and intensive proliferation of the tumor cells [67]. Apparently no other protein can replace the histone. It is known, for example, that the protein bound with the DNA of E. coli (2.3:1) inhibits the template activity of DNA in RNA synthesis less efficiently than the histones of higher organisms [68]. Exogenous histones do not penetrate into the nuclei of cultured mammalian leucocytes even after 24 hours of incubation and remain mainly on the cell surface [69]. The specificity of interaction between histones and nucleic acids can be determined from the efficiency of sedimentation of the complex [70]. According to one view, substitution of aspartic acid with valine in the histone makes the latter incapable of repressing DNA synthesis, and this is one of the causes of the cancerous proliferation of the cells [71].

Incorporation of tritium-labeled thymidine into chromosomes may take place during the G_1 period by nonconservative repair synthesis of single DNA strands which have been depolymerized at the sites of mRNA synthesis. The template activity of DNA during this period can be gauged from the rate of this incorporation.   The template activity of DNA becomes especially pronounced toward the end of the G_1 period, when the genes controlling the release of enzymes for nucleic acid synthesis as well as the genes responsible for the release of one of the histone fractions are in an active state [72].

It is not yet known which of the two chromonemata is functional during the G_1 period — the dextrorse, the sinistrorse, or perhaps both.   At any rate, the functional part of the chromosome consists of euchromatin, which accounts for 20% of the total chromatin of thymus cells and as much as 40% in the highly differentiated liver cells.   What is the composition of the heterochromatin — the functionally inert deoxyribonucleoprotein?   Assuming that the chromonemata are isologous and that the sinistrorse α-helix is less stable than the dextrorse one, it is conceivable that only the dextrorse chromonema (40% of its chromatin) is functional during the G_1 period, whereas the entire sinistrorse chromonema and the remaining 60% of the dextrorse one (in all, 80% of the total chromatin) are repressed by the histone during this period and form masses of heterochromatin granules near the inner surface of the nuclear membrane.   DNA synthesis begins by the end of the G_1 period when the cell has attained a critical mass [73].

The chromosome during the S period of interphase.   Replication of chromosomes during the S period involves a semiconservative (template) synthesis of DNA.   This process is conveniently studied by autoradiographic analysis of metaphase chromosomes containing tritium-labeled thymidine incorporated during the synthesis of daughter DNA strands.   Although the objects of autoradiography are dense chromosomes fixed during metaphase, the distribution of grains in the β-sensitive emulsion may reveal the fate of the chromosomes during the S period.   Not surprisingly, the most detailed information on the mechanism of chromosomal replication has been obtained from simple objects such as phages and bacteria, the chromosome of which contains a single circular DNA molecule.

Cairns [28] has proposed a model for replication of the circular chromosome on the basis of autoradiographic data.   This model is based on the concept of the flexible shaft, proposed to explain replication of the linear DNA molecule [74].   The linear double helix unwinds as it rotates like the flexible shaft of a speedometer at the speed necessary for its replication within a given period of time.   The torsional force necessary for this rotation exerts a minor additional stress on the chemical and hydrogen bonds but does not sever them.   It has been calculated that if unwinding of each DNA coil around its axis causes rotation of the whole bulk of the double helix, replication of even a large phage, with a DNA molecule of $6 \cdot 10^4$ nm long, would consume only 0.15 cal per mole of polynucleotide unit or about 1% of the energy stored in the energy-rich nucleotide triphosphate bond. According to Cairns, the junction of the ends of the circular chromosome serves as a molecular swivel which rapidly uncoils the DNA of E . coli at the growing point.   He calculates that the rolling circle must rotate around its axis at a rate of about 700 revolutions per second in order to unravel 10^6 turns of the DNA spiral in 30 minutes, as is the case in E . coli.

This model has found its way into some textbooks [29, 75], but it apparently ignores some of the advantages of the circular form of DNA.   In particular, the idea of a flexible shaft with a molecular swivel appears rather farfetched.   On the other hand, the ideas of Gamov and Bloch on the topological aspects of the semiconservative replication of DNA [76, 77] assume a profound significance with respect to circular DNA, especially in the complex chromosome of multicellular organisms, consisting of two or more genonemata.   According to Gamov, the plectonemic DNA double helix can segregate and replicate without breaks or other complications if its replication is geared to formation of a paranemic double helix composed of two daughter spiral helices (Figure 5.3).   Bloch assumes that formation of the two new double helices is accompanied by uncoiling of the parental double helix.   This is preceded by cleavage of the hydrogen bonds at the growing point and rotation of the bases by about 180° around the single bonds with the sugar moeity since only such an arrangement allows the free triphosphate bases to find their complementary place in the newly synthesized DNA strand.   In this manner the dextrorse α-spiral yields two sinistrorse α-spirals and the sinistrorse spiral two dextrorse ones.   Since histone participates in this complex process, its formation must proceed simultaneously with replication of DNA.

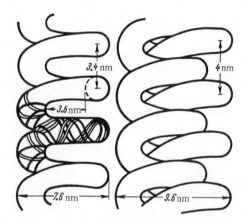

FIGURE 5.3.   Model of the tertiary structure of DNA in the form of a regular spiral helix (left) and a paranemic spiral helix (right), resulting from the replication of two spiral DNA helices

A model for the replication of circular DNA can be proposed on the basis of the ideas of Gamov and Bloch without assuming the existence of a molecular swivel or the rotation of DNA like a flexible shaft (Figure 5.4). According to this model, unraveling of the circular plectonemic DNA helix begins at a single growing point which moves along the perimeter of the circle as in the Cairns model.

However, unwinding of the strands of the DNA double helix takes place near the growing point and involves rotation of the bases by 180°.   This does not necessitate rotation of the whole helix from the starting point to

the growing point, nor does it require a molecular swivel. The unwound
DNA strands serve as templates for the synthesis of daughter strands at
the growing points and thus yield two sinistrorse α-helices. These, according to Gamov, fit closely into a paranemic double helix composed of spiral
helices. If unwinding is to proceed without breaks, the turns of each spiral
helix must be 3.4 nm apart. Formation of the paranemic helix requires
that the diameter of the primary helix be decreased by about 0.6 nm as a
result of tighter coiling, or that the distance between the turns of the
secondary helix be increased by the same length (that is, to 4.0 nm instead
of 3.4 nm). If the DNA of biological systems exists for the most part as a
nucleohistone complex, it may be assumed that removal of histone from the
parental helix and covering of the daughter helices with histone proceed
simultaneously with movement of the growing point along the perimeter of
the circle.

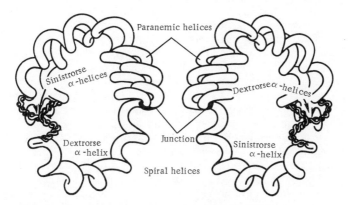

FIGURE 5.4. Molecular model for replication of the circular chromosome involving formation of daughter spiral helices and their arrangement in a paranemic
helix as proposed by Gamov [76] and the possible transition of the dextrorse
α-helix into a sinistrorse one and vice versa according to Bloch [77]. The model
demonstrates how rotation of the bases by 180° around the single N-glycoside
bonds yields two sinistrorse α-helices from the dextrorse α-helix and two
dextrorse α-helices from the sinistrorse α-helix.

Thus, the growing point moves along, leaving in its wake a developing
paranemic helix composed of two daughter α-helices which are mirror
images of the parental helix. The diameter of the nucleohistone turns of
the secondary helix may reach 10 nm and that of the double coils of the
paranemic helix 12–15 nm. The mirror-image twin helices can be converted by a similar mechanism into exact copies of the parental spiral,
either immediately after replication or some time later. This process
may be repeated a number of times during the cell cycle, according to the
requirements of the replication program.
    The swivel-less model shown in Figure 5.4 eliminates many of the
topological, energetic and biochemical obstacles which Cairns' swivel model
cannot overcome. Its main advantage is that it rules out the risky uncoiling

of DNA to single stranded areas in the vicinity of the growing point. The rotary degrees of freedom for the uncoiling parental DNA molecule and the coiling daughter molecules are confined to the immediate vicinity of the growing point of the DNA. The energy of the rotary movement is not dissipated as heat in a swivel or in friction between a flexible shaft and the viscous plasma; instead, it is directed almost entirely into the forceful winding of the plectonemic daughter helices and their arrangement into a paranemic helix. If the spacing of the secondary DNA helix equals that of the primary Watson-Crick helix (3.4 nm), all the torsional movements of the parental and daughter helices can be so balanced as to avoid dissipation of the hydrogen bond energy (structuralization energy) into heat, and the entropy of the system will not increase substantially throughout the process. In other words, the energy spent on structuralization of the parental helix can be transformed almost entirely into the structuralization energy of the daughter helices; all this is in accordance with the fundamental thermodynamic principle of living systems, established by evolution, namely, the maintenance of entropy at a quite low constant level [78].

The closed DNA circle represents a fairly advanced elementary open system: energy arrives at the growing point together with the deoxyribonucleotide triphosphates within the energy-rich bonds of these molecules, at the rate of 12.5 cal/mole per dephosphorylation (and polymerization) event, i. e., in a strictly quantized manner. The start and rate of the complementary polymerization of DNA are largely controlled by the DNA itself: they depend on its condition, coiling and proteinization. Owing to the high rate of DNA synthesis and the "pumping in" of nucleotides for polymerization, a concentration gradient of nucleotides develops around the growing point; this gradient is directed toward the growing point and shows a peculiar "cyclonic" convergence from the periphery toward the center. An outflow of phosphate residues takes place in the opposite direction. The inhibitory effect of exogenous phosphates on the synthesis of DNA can be attributed to the disappearance of the concentration gradient, bringing the outflow of endogenous phosphates and the corresponding inflow of nucleotide triphosphates to a halt.

The currently postulated mechanisms of DNA replication [1, 79—81] apply equally to cyclic and linear structures. In particular this is true of the hypotheses regarding the quantum nature of the specific collective forces operating during the convariant replication of elementary cell structures [3, 82], the energetic and kinetic factors governing selection of complements from the pool of the four free nucleotides [83], the quantum chemical factors involved in polycondensation of nucleotides and the energetic aspects of the rupture of the hydrogen bonds of the double parental spiral in the semiconservative mechanism of replication [85].

Despite the evidence obtained regarding the replicon organization of the chromosome and the multisite nature of its replication [9, 86], it appears that replication of the chromosome of multicellular organisms is based essentially on the above principle of replication of the circular DNA macromolecule. Regardless of the manner of chromonemal replication — by replicons or immediately throughout the ring — it appears that the beginning of replication depends entirely on the structural and metabolic conditions in the vicinity of the growing point.   Each replicon can include two growing points.   On the basis of an autoradiographic study of chromosomal replication in rootlets of B e l l a v a l i a, Taylor [9] arrived at three important conclusions: 1) the subunits of the chromatid are not similar to one another; 2) the marked differences in distribution of the label in the sister chromatids result from asynchronous replication of two subunits of the chromatid; 3) an exchange apparently takes place during DNA replication between similar (isologous and homologous) subunits of the chromatid. These conclusions have been largely confirmed in many laboratories [6, 7] in experiments with various groups of higher organisms [87, 88], and also by Taylor himself [33] with Chinese hamster cells.   This last publication stresses the low frequency of exchange between sister chromatids when there is a very large number of replicons, the independent replication of which is preceded by frequent single breaks of the DNA chain and ends with their reunion.

Observations of the postradiation rearrangements of the chromosomes of C r e p i s ' c a p i l l a r i s through a series of nuclear cycles and mitosis (polyploidy) [89] have confirmed Taylor's conclusions regarding the structure and replication of chromosomes.   Study of the second mitosis following irradiation revealed that the chromatid consists of at least two subunits capable of replication and further rearrangements — isochemichromatid deletions.   The dissimilarity of the chromonemata is evident from the appearance of about 20% chromatid-type rearrangements following irradiation of seeds of C r e p i s   c a p i l l a r i s during the G_1 stage, when only chromosome-type rearrangements might be expected.   Assuming that the chromonemata (half chromatids) are dissimilar, the theoretical yield of chromatid rearrangements upon irradiation during the G_1 stage is 25%.   The appearance of 50% chromatid rearrangements when the seeds are irradiated with neutrons [90] can be attributed to dissimilarity of chromonemata.   Irradiation of the cells during the S stage causes not only chromosome and chromatid rearrangements but also, in some cases, chromosome-chromatid rearrangements within the same chromosome [89].   This can be explained by an asynchronous synthesis of chromosomal subunits.   The observed dissimilarity of the chromonemata may be explained on the grounds that they consist essentially of two different α-helices — one dextrorse and one sinistrorse.

According to autoradiographic data [36], the asynchrony of DNA synthesis in the chromosomes of C r e p i s   c a p i l l a r i s and T r a d e s c a n t i a results from their length: replication of DNA begins simultaneously in the euchromatin regions of all chromosomes of the set and ends in the heterochromatin regions at different times according to their length.   The rate of this synthesis is about 1,500 nm per hour.   The asynchrony of DNA synthesis in the euchromatin and heterochromatin may indicate asynchronous replication of the dextrorse and sinistrorse helices.   Although it is difficult to determine which chromonema — the dextrorse (α) or the sinistrorse (α') — replicates first, the replication process as a whole may be represented by the following formula:

$$\alpha + \alpha_1 \xrightarrow{\text{start of } S} 3\alpha' \text{ or } 3\alpha \xrightarrow{\text{end of } S} 2\alpha + 2\alpha' \xrightarrow{G_2\text{-stage}} 2(\alpha + \alpha') \,.$$

Depending on whether the dextrorse or sinistrorse chromonema replicates first, the chromosome will contain either three sinistrorse or three dextrorse chromonemata during the S stage. The sinistrorse helices are considered to be less stable [91]. The S stage possibly begins with transformation of one dextrorse chromonema into two sinistrorse ones. Conceivably, replication of one of the newly formed chromonemata at the end of the S stage is twice as probable as replication of the "old" chromonema. Thus, autoradiography during the first and second mitoses may reveal half-labeled chromonemata as well as unlabeled ones [9].

The nonarticulated model of chromonemal replication provides an acceptable solution to the problem of DNA polymerase. It is known that this enzyme operates only in one direction, from $3'OH$ to $5'P_2$ and not the other way round. The hypothesis that two different DNA polymerases exist is yet to be confirmed.* Assuming a single DNA polymerase, the swivel model overcomes this difficulty by postulating alternation of the template, i. e., that the enzyme changes the template at each step of the process and incorporates nucleotide monophosphates simultaneously in both chains with the normal sequence and antisequence of sugar-phosphates. With the nonarticulated model, however, matters are simpler — both extending chains at the growing point (see Figure 5.4) can be viewed as a single chain with the sequence $3'OH \rightarrow 5'P$ and the presumed breaks in the circular DNA molecule (owing to the shift of the initial points of synthesis on either DNA chain by four nucleotides [92]) are not obligatory.

Autoradiography [9] reveals occurrence of intra- and interchromosomal exchanges during the S stage: the second mitosis following addition of tritium-labeled thymidine yields chromosomes composed of labeled and unlabeled parts. It is easy to imagine an intrachromosomal exchange between three isologous chromonemata at a single centromere, where an exchange of parts can take place between the newly synthesized labeled chromonema and the old unlabeled one. An interchromosomal exchange is more difficult to visualize. A somatic crossover apparently requires the same conditions as the meiotic one — namely, mutual attraction between homologous chromosomes in the nucleus during the S stage, conjugation, and segregation during the G_2 stage. In principle, this is quite possible. The forces of attraction between homologous chromosomes, which can be regarded as elongate and on the whole neutral molecules, possibly result from interaction of electromagnetic fields created by zero-point fluctuations in the electric charge density along the eu- and heterochromatin chains [93, 94]. These forces of attraction are enhanced if the chromosomes lose their neutrality as a result of two accidentally opposite transitions of the helices during replication — namely, $\alpha \rightarrow \alpha'$ in one of them and $\alpha' \rightarrow \alpha$ in the other. In this case one of the homologous chromosomes will consist of three sinistrorse helices, and the other of three dextrorse ones. These antihelices (antisolenoids) can apparently create a force of attraction with a range much greater than their own length. It may be assumed that conjugation, synopsis and segregation of the chromosomes during the S stage of mitosis proceed in the same manner as during the first division of meiosis.

---

* In a recent publication, Knippers (Nature, 228:1050. 1970) reported the presence of two DNA polymerases in the cell — replicational and reparative — which differ in their activity and position in the nucleus.— The Editor.

The chromosome during the G_2 stage of interphase.   The postsynthetic G_2 stage can be regarded as the beginning of distribution of the newly synthesized chromosomes into two isologous sets: the two paranemic double helices $2\alpha + 2\alpha'$ of the late S stage begin to form two chromatids composed of dextrorse α and sinistrorse α' chromonemata during the G_2 stage, that is, paranemic helices and redistribution of the chromonemata.   The synthesis of DNA almost ceases during this period whereas the synthesis of histone continues at a high rate throughout the G_2 stage [36].   Incorporation of a small amount of tritium-labeled thymidine can be attributed to continuing (or incomplete) intra- and interchromosomal exchanges requiring repair synthesis of DNA at the points of cleavage near the conjugating sections of the chains.

The chromosome during the prophase of mitosis.   During prophase the chromonemata in the chromosomal chromatids begin to coil with the participation of a new structure developing at this time — namely, the mitotic apparatus of collagen spindle fibers extending from the chromosomes to the two poles or centrioles.   The condensation of the chromosomes during prophase after attachment of the spindle fibers to the centromeres can be outlined as follows.   1) Union of the looped ends of the dextrorse and sinistrorse chromonemata into two isologous sister chromatids (formation of the telomere).   2) The tension of the spindle fibers extending to the two opposite poles causes a rotary movement of the kinetochores of centromeres, while the terminally attached (circular) chromonemal loops become tightly twisted.   3) The tightly twisted chromonemata form fifth-order helices (about 200 nm in diameter) so that their ends gradually advance toward the centromere (almost perpendicularly to the stretched fiber of the spindle).   At this stage of coiling the chromosomes enter into metaphase.

A further condensation of chromatin during metaphase is brought about by the orderly union of the fifth-order chromonemal antihelices into a single chromatid cylinder.   A pair of such cylinders form the metaphase chromosome.   Distribution of the metaphase chromosomes in the equatorial plane appears to be the result of rather rapid rotation of the spindle around an axis passing through the centriole.   Because of this rotation, the largest chromosomes lie closer to the periphery of the equatorial plane of the spindle.

Thus the mitotic cycle is repeated: the tension of the spindle fibers in early anaphase causes a division of the centromeres and chromatids, while continuing rotation of the kinetochores during late anaphase leads to further condensation of the separated chromatids.   Disappearance of the spindle is accompanied by uncoiling (backward development) of the chromosomes.   All these stages of coiling and uncoiling can be satisfactorily simulated in mechanical models of chromosomes incorporating imitation kinetochores (centromeres) which rotate when tension is applied to the spindle fibers.

Thus, the whole life cycle of the cell, i. e., the cycle of coiling and uncoiling of its chromosomes, can be accurately timed by cytomorphological electron microscopic and autoradiographic procedures.   Figure 5.5 shows an example of such timing with schematic designations of the condition of the chromosomal DNA during the consecutive periods of interkinesis and the various mitotic stages (cited from [95]); the time scale selected for the mitosis is about 20–24 times larger than that of interkinesis, but within interphase the G_1, S and G_2 stages are represented to scale. Assuming that the average life cycle of the cell is 24 hours, the respective durations of the stages are as follows: $G_1 \approx 14$ hours, $S \approx 5$ hours, $G_2 \approx 4$ hours, and $M \approx 1$ hour.

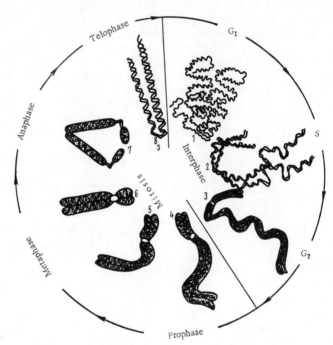

FIGURE 5.5. The state of chromosomal DNA during different periods of inter-kinesis (1–3) and mitosis (4–8):

1) G_1 stage. The chromonemata are uncoiled and the DNA partly deprotein-ated (at the sites of synthesis of mRNA); 2) S stage. DNA replication is in progress, the replicating part being in a strongly hydrated state (in the form of "nucleoplasm"); 3) G_2 stage. Distribution of the isologous daughter spiral nucleohistone helices into two isologous sets of sister chromatids; 4) prophase. Further twisting of the distributed helices in the chromonemata (supercoiling) and union into chromatids; 5) early metaphase. Distribution of the chromo-somes in the "equatorial" plane of the nucleus; 6) metaphase chromosome composed of two sister chromatids; 7) anaphase chromosomes (diverging sister chromatids); 8) telophase. Uncoiling (backward development) of the chromosomes.

Chromosomes of nondividing cells. Advances in knowledge of the fine structure of the chromosomes of dividing cells during the G stage of inter-phase have enabled an equally successful analysis of the structural and functional organization of chromosomes in nondividing cells. It was found not long ago that cell differentiation occurs by means of a single replicon expansion of different parts of the uncoiled chromosomes which vary from one organ or tissue to another, and that this process apparently involves multiple replication of DNA in the individual replicons. The chromosomes of differential cells contain a complete assortment of genes, many of which are in a latent state — repressed and not replicated. Hence the division of labor among the different cells of a particular organ (e. g., the liver) or between different organs (e. g., the bone marrow and intestinal epithelium) in the production of proteins and other compounds essential to life.

In a sense, resting cells can be defined as interphase cells with G_1 stage of indefinite duration. Under suitable conditions these cells can emerge from the resting state and begin to divide, as is the case with liver cells 7 hours after hepatectomy.

A special case of the single replicon expansion of chromosomes is polyteny. The giant polytene chromosomes of dipterans arise by endomitosis — division of the chromonemata without condensation by coiling and distribution into daughter sets [6, 7, 12, 13, 75, 95, 96]. The chromonema of the normal chromosome, which replicates more than 1,000 times serves as the template [12, 95]. Each polytene chromosome, e. g., those found in the cells of the salivary glands of Drosophila, contains bundles of about 1,000 chromonemata 100—200 times as long as the normal coiled metaphase chromonema [95] but equal to the interphase length of the latter. The discs of genetically inactive heterochromatin can be regarded as masses of chromonemal zones fused side by side. The concentration of the discs in the distal part and their spacing out in the proximal part indicate a steeper coiling near the centromere in comparison with the ends of the chromosome. The intervals between the discs contain euchromatin, which is poorly visible under the ordinary microscope and appears to be composed of the nucleo-histone fibrils of the spiral helices. The deproteinated and uncoiled areas of the latter represent actively functioning gene loci.

Not surprisingly, the centromere and genetic loci of the arms of the polytene chromosomes lie in the same order as in the normal chromosomes; indeed, the chromonemata of the polytene chromosome, arranged side by side, evidently correspond to the same elementary chromosomal loop with exactly the same (isologous) set of genes as in the normal chromosome, their sole distinction being that they are possibly coiled according to a different program. In the case of polyteny, this program does not include formation of a DNA superhelix.

Thus, the body of the polytene chromosome consists of a bundle of isologous chromonemata, not coiled right to the end, containing over 1,000 chromonemal loops stretched and twisted into plectonemic helices with each coil corresponding to one gene. This bundle results from limitations in the program for the supercoiling of DNA, somatic crossing over and cell division, despite normal replication and coiling during endomitosis. One possible reason for the absence of supercoiling of DNA during endomitosis is the noncyclic form of the chromonema.

The chromosomes during meiosis. Formation of the haploid sex cells (gametes) in the genital glands (gonads) from diploid somatic cells of these organs (oogonia in the female, spermatogonia in the male) involves a reduction division called meiosis. This important nuclear process has been thoroughly studied since its discovery at the end of the last century [6, 7, 10, 12, 13, 75, 78, 95, 96]. In meiosis, doubling of the chromosomes is followed by two consecutive divisions of the nucleus, in contrast to mitosis in which alternation of chromosomal replication and cell division ensures a constant diploid set of chromosomes. As a result of meiosis, the diploid cells (oogonia or spermatogonia) yield four gametes with different haploid sets of chromosomes. Figure 5.6 shows diagrammatically the process of meiosis [7] with particular reference to the state of the chromosomes during the five stages of prophase in the first division. The prophase of the first meiosis is more prolonged than the mitotic prophase, and its various stages include chromosomal transformations which do not exist in mitosis.

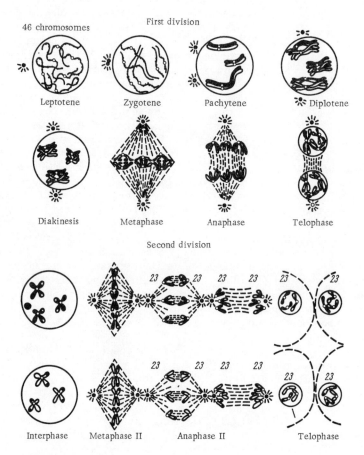

FIGURE 5.6. Scheme of meiosis [7]. Two consecutive divisions yield four gametes with different haploid sets of chromosomes. The figures indicate the number of chromosomes at the corresponding stages of the first and second divisions.

During the leptotene stage (early prophase) the chromosomes take the form of slender, long, solitary, slightly coiled filaments. In the zygotene stage, the homologous filaments become closely paired throughout their length and conjugate with one another by a process termed synapsis. The high accuracy of the conjugation is evident from the fact that synapsis involves only homologous chromosomes and takes place at strictly determined points on the homologs. The physical nature of attraction between these identical structures remains obscure. One possible explanation is that the chromosomes contain only one doubled interphase chromonema at the beginning of prophase of the first meiosis, i.e., they contain only three dextrorse or three sinistrorse chromonemata. Conceivably, the uncoiled chromosomes — that is, the elementary chromosomal filaments in the form of dextrorse and sinistrorse spiral helices — can attract one another like solenoids of different polarity. This would mean that at the zygotene stage the cells contain antihelical (mutually attracting) homologs. According to one view, synapsis proceeds with a zipper-like action until the two homologs become intimately joined into a bivalent [97]. Microscopical examination reveals replication of the homologous chromosomes, taking place some time after formation of the bivalent. This replication can be interpreted as delayed transformation of one dextrorse chromonema into two sinistrorse ones and of one sinistrorse chromonema into two dextrorse ones. As a result, the bivalent consists of four pairs of dextrorse and sinistrorse chromatids — that is, $4(\alpha + \alpha')$. Thus the zygotene chromosomes are well prepared for the supercoiling which takes place during the next stage.

In the pachytene stage, the chromosomes are in the form of thick helices; the homologs are so intimately joined that it is very difficult to distinguish the two chromosomes. During the next stage, diplotene, the replicated homologs separate from one another. Each of the chromosome pairs (bivalents) forming the synapsis consists of two chromatids arranged in a tetrad. The points of contact between nonsister chromatids of the same tetrad are termed chiasmata. Each tetrad usually bears at least one chiasma. The chiasmata correspond to the sites of breakage and exchange of material between two homologous chromosomes during the zygotene stage in a process known as crossing over. In other words, chiasmata at the diplotene stage are evidence that crossing over has taken place during the zygotene stage.

During diakinesis there is additional condensation of chromatin (supercoiling) and displacement of the chiasmata toward the ends of the chromosomes (terminalization). At the end of diakinesis the nuclear membrane disappears and the bivalents with their maximally coiled chromosomes move to the equatorial plane of the first meiotic division.

In the first meiotic metaphase, the homologous chromosomes forming the bivalent lie in the equatorial plane, while in anaphase they segregate and move toward opposite poles. During telophase, haploid sets of chromosomes congregate at the poles and these form the interphase nucleus prior to the second meiotic division.

The second meiotic division proceeds in the same way as mitosis: the chromatids of each chromosome of the haploid set segregate and move to opposite poles of the cell. As a result of the crossing over, each chromosome of the gamete (one linkage group) contains elements from both parental chromosomes. In the ideal case of an exchange of genetic material by crossing over, the mixed segments of the chromatids (gametal chromosomes) are reciprocal (that is, strictly complementary, mutual and identical). In some cases, however, especially as a result of various physical and chemical interventions, the exchange is nonreciprocal (nonmutual and nonidentical) and leads to all kinds of chromosomal mutations.

The chromosomes of specialized cells. Although the existence of a double sinistrorse α-helix of DNA was postulated long ago [77, 91], no such helix has yet been found. However, the existing micrographs and some electron micrographs provide indirect evidence of a sinistrorse coiling [8, 9, 98, 99].

Of particular interest in this respect are the chromosomes of specialized cells such as spermatozoa. It is known that the head of the spermatozoid consists almost entirely of DNP. Analysis of the protein fractions isolated from hog spermatozoa shows the absence of histones and protamines [100]. The protein of hog spermatozoa is related to the basic keratins and has a very high arginine content. This protein is apparently better able to protect the DNA during its transfer to the ovicell than histones.

In a study of the chromatin of the head of spermatozoa of differing origin, N. K. Kol'tsov [12] was able to make microscopical drawings showing the nature of the superhelices as long ago as 1908. He found spermatozoa with sinistrorse superhelices among those with dextrorse ones.

The properties of the chromatin of the superhelices are better known today than 60 years ago. Judging from such factors as length and thickness of the continuous cords, their behavior on staining and treatment with different chemical agents and the effects of high osmotic pressure inside the spiral "ampule," these helices appear to be composed of deoxyribonucleoprotein and equipped with a complete haploid set of male chromosomes. The universal collinearity of DNA and the protein product suggests that the spermatozoa with sinistrorse DNA superhelices form complexes with sinistrorse protein helices. If the sinistrorse nature of the protein has a phenotypic manifestation in the micro- or macroworld and is inherited in accordance with Mendel's laws [101], one might expect the existence of a sinistrorse genotype and "anticode," as postulated by Zachau [102]. Although sinistrorse protein helices may arise from dextrorse ones by rotation of the amino acids around certain bonds [91], a genetically determined production of sinistrorse protein would be far more efficient.

Today the existence of sinistrorse DNA molecules can be established in a variety of ways, including electron microscopy, which reveals the Watson-Crick helix and enables determination of the distance between its coils and the presence of protein-free and protein-bound areas [103]. The diffraction pattern can hardly be expected to show the rare and relatively unstable sinistrorse α-helix of DNA in the presence of the more common dextrorse α-helix. Tikhonenko established the existence of sinistrorse superhelices in DNA-containing viruses (Molekulyarnaya Biologiya Virusov, p. 86. "Nauka." Moskva. 1971).

Finally, a few more comments on the postulated looped structure of the chromosomal subunits. The mechanism of crossing over and its genetic consequences — chromosomal mutations — can be convincingly explained in terms of the classical hypothesis of the linear arrangement of genes in the chromosomes of multicellular organisms [104—106]. Since the requirement for linearity can only be met by a single stranded chromosome composed of one double helix and is not satisfied by a two- or three-dimensional chromosome, the existence of a two-dimensional looped chromonema needs an explanation. This problem can be solved by comparing the classical and modern rules for preparing genetic maps of chromosomes and also the results of their application. At the beginning of this century Morgan and his co-workers concluded quite logically that frequency of rupture of the bonds between genes linked in a single group is proportional to the distance between the genes. An accidental break in the interval between two genes becomes more probable the further apart the genes; the distance between genes which are separated in 20% of the cases is 4 times greater than that between genes separated in 5% of the cases [12]. The number of ruptures of the normal link between the genes of the linear chromosome was taken as the unit of distance between them and was called a morgan. The diagrams showing the relative (collinear) arrangement of the genes in a particular linkage group (chromosome) are referred to as genetic maps.

The linear concept of the genetic map has enabled a detailed analysis of the position of the genes in the linkage groups of a number of organisms including Drosophila, maize, tomato, Neurospora, E. coli, etc. [75, 95]. Since the crossing over between the farthest points of a chromosome cannot exceed 50%, a linear genetic map of the chromosome based on the frequency of crossing over can contain no more than 50 morgans (crossover units), or even less than this number since all chromosomes include large areas of genetically inert heterochromatin. In fact, however, the genetic maps of some Drosophila chromosomes contain about 100 morgans (second and third linkage groups), and some chromosomes of maize as many as 120—160 morgans (first, second and third linkage groups), i.e., 2—3 times the expected values. This discrepancy is usually explained by assuming that the sum total of crossover units on the map comprises events located in overlapping parts of the chromosome. Such an interpretation, however, can hardly explain a discrepancy of this magnitude (2—3 times or more). The question inevitably arises whether the discrepancy between the map length (100—150 morgans) and the number of recombining loci of the diheterozygote (50%) stems from the principle of linearity in mapping of the chromosomes of multicellular organisms. In other words, the problem can perhaps be solved by a different mapping procedure.

Molecular genetics often resorts to circular genetic maps as the only maps possible. In a study of the fine structure of several genetic loci of Neurospora by interallelic complementation, Giles [107] concluded that although most complementation maps of this and other organisms are apparently linear, an increasing body of evidence demonstrates the necessity of circular maps. The recombination and complementation data for one of the Nuerospora loci were interpreted in terms of a circular map. Giles believes that this map of the locus fits the genetic map if the latter has the form of a double helix; in this case the mutant sites of the two maps will coincide. In other words, analysis of the fine structure of the gene leads to the conclusion that the gene locus possibly represents a single twist of the chromonemal loop, bordered marginally by two heterochromatin junctions (see Figure 5.2).

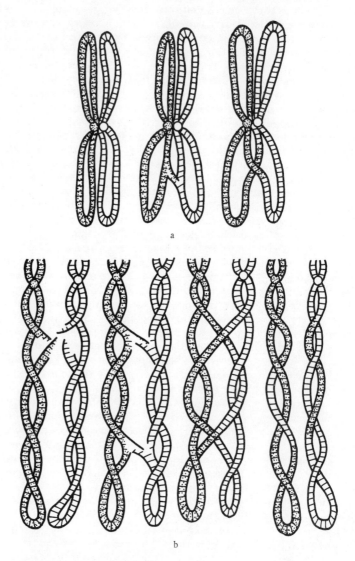

FIGURE 5.7. Two possible molecular mechanisms of crossing over:

a) consecutive crossing over at a primary single break of a looped DNA chain during intrachromosomal exchange; b) crossing over by pairing of complementary ends of DNA formed by double oblique breaks (caused by endogenous enzymes or external agents). Spiral helices are portrayed arbitrarily in both cases.

The looped model of the chromonema provides the same tentative explanation of the molecular basis of synapsis, intra- and interchromosomal reciprocal exchanges during crossing over and chiasmata and their terminalization as does the linear model. The molecular mechanism of crossing over is shown diagrammatically in Figure 5.7. This mechanism is based on the complementarity of conjugating sections of DNA at the breakage site of the polynucleotide chains. Breaks occur at specific sites, possibly with the participation of DNase. Restoration of the DNA chains involves dark repair enzymes. As yet little is known of the nature of these enzyme-induced breaks. It may be assumed, however, that cleavage of the DNA chain does not involve the loss of even a single nucleotide at the site of the enzymatic attack, i.e., it occurs strictly at a phosphodiester bond (most probably $C(3') = O = P$) and not between carbon atoms, $C(3') - C(4') - C(5')$. In some cases the loss of a single nucleotide out of a total of 5,500 (as in the phage $\varphi X174$) totally destroys the functional activity [30]. A complete exchange requires a double break in the polynucleotide chains; this may occur, however, either consecutively or simultaneously, as shown in Figure 5.7. In the case of double oblique breaks conjugation requires single strand complementary ends of DNA. With respect to phage genomes it has been shown that closure of the linear DNA into a ring takes place in the presence of free single strand ends bearing about twenty complementary nucleotides [108]. A similar length of single strand complementary ends of DNA may be necessary for the union of chromosomal subunits in multicellular organisms.

Collinearity of the genes in the normal and polytene chromosomes of dipterans was stressed above. This collinearity is not upset during the meiotic recombination of genetic material even in the case of nonidentical and nonreciprocal exchanges, that is, chromosomal mutations. The mutations of polytene chromosomes are therefore not surprising: they reflect rearrangements of the normal chromosomes, determining all the phenotypic characters.

Genetic analysis does not enable more accurate detection of chromosomal mutations than cytogenetic analysis, which can reveal such phenomena as deletion or doubling of heterochromatin discs in polytene chromosomes, changes in the position of the different discs, or reversals of their normal sequence. These defects persist because they do not impede the normal course of mitosis, meiosis and endomitosis. Their formation involves the same repair enzymes as those participating in the reciprocal exchanges which do not cause mutations. The looped model of the chromosome does not introduce any additional difficulties in interpretation of the results of nonreciprocal exchanges. On the contrary, some mysterious properties of chromosomal mutations can be plausibly explained on this basis. It is known, for example, that terminal inversions are extremely rare among chromosomal mutations, whether spontaneous or induced by some agent. This can be attributed to the nonadhesiveness of the looped telomeric end of the chromonema. The ability of chromosomal mutations to transform genes from bipolar to unipolar may be similarly explained: this transition possibly involves formation of a molecular "stump" comparable to a telomere, i.e., a new loop closed at the end. Absence of back mutations in this case can be explained in the same manner as the lack of terminal inversions.

To sum up, the available cytomorphological and electron microscopic data on the structure of the chromosome and its subunits in multicellular organisms, as well as the biochemical and autoradiographic evidence as regards the function of the chromosome during the G_1 and S stages and the genetic and cytogenetic data on linear arrangement of genes along the chromosomes and exchange of chromosomal material by crossing over, all yield the general impression that the structural and functional organization of the chromosome of multicellular organisms corresponds to a progressive coiling of dextrorse and sinistrorse double α-helices of DNA in looped chromonemata. This model differs from the preceding ones in its dynamism; moreover, it allows a follow-up of the fate of the chromosome throughout the life cycle of mitotic cells as well as during the different stages of meiosis. The vast and specific program carried out by chromosomes during the meiosis and mitosis of sex and somatic cells, both dividing and non-dividing, can be easily visualized on the basis of the proposed model. In addition, this model can also explain the function of euchromatin (see Figure 5.2) and the polyreplicon or single-replicon replication of the DNA template (see Figure 5.4).

RADIATION-INDUCED DAMAGE TO NUCLEIC ACIDS IN THE LIVING CELL

The genetic role of the chromosomal DNA of the cell as a template is evident from the preceding discussion: the template mechanism of the information transfer from DNA to protein by means of the genetic code ensures specificity of protein biosynthesis in the particular cell, while the template mechanism of complementary polymerization of DNA by replication is responsible for specificity of the same biosynthesis throughout the cell population. The structural and mechanical function of DNA in mitosis and meiosis during differentiation and specialization of cells — that is, its ability to coil and uncoil, conjugate and exchange fragments, pass intact from maternal to daughter somatic cells, and undergo reduction during its transfer from the diploid sex cells to the haploid gametes — is similarly evident.

Radiation-induced damage to DNA in the cell can obviously affect its genetic, structural and mechanical functions in the chromosomes. However, no one has yet seen a lesion of DNA in cell chromosomes as a result of ionizing radiations. Attempts have been made to simulate such lesions qualitatively by irradiation of nucleoprotein complexes isolated from cells, and also viruses and phages, but the results have been disappointing. The main difficulty in the in vivo analysis of radiation-induced damage to chromosomes lies in the heterogeneity of DNA.

Because of its heterogeneity, chromosomal DNA differs markedly from the DNA used in radiochemical models as a substrate of radiobiological reactions. As we noted in describing the structural and functional organization of the chromosomes of multicellular organisms, the state of aggregation, condensation and proteination of their DNA vary during the life cycle

of the cells.   Extraction of DNA from the cells of an organ or tissue inevitably yields a mixture of macromolecules comprising fractions in different states of aggregation, protein-bound and protein-free, uncoiled and
coiled into superhelical supramolecular structures.

The number of lesions in the macromolecules isolated from irradiated
asynchronous cell populations varies from one fraction to another.

Moreover, cellular DNA has a rather wide discrete spectrum of molecular weights since the numbers of chromosomes of different lengths and
the arm length of chromosomes vary.

An example of this is the spectrum of molecular weights of DNA in human
chromosomes [7].   Such a spectrum can be obtained on the basis of the
absolute length L_a (nm) and the relative length L_r of the chromosomes (the
latter being expressed as a percentage of the total length).   The average
absolute length of the chromosome determined from this spectrum is
$L_a = 6,670$ nm.   Such a chromosome must contain an average of about
10^6 nucleotides since the nucleotide content of all 46 chromosomes is about
10^4 times greater than that of phage φX174 DNA, which consists of $5.5 \cdot 10^3$
nucleotides [30].   This corresponds to the number of genes in human cells
$(2.3 \cdot 10^4)$.

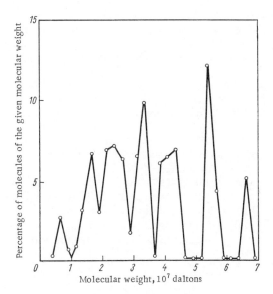

FIGURE 5.8. Approximate spectrum of molecular weights
of human chromosomes

In accordance with Figure 5.1, the subunit containing one DNA molecule
corresponds to half the chromonema found in one arm.   Each chromosome
contains 8 such half chromonemata, which means that each average molecular subunit comprises about $1.25 \cdot 10^5$ nucleotides and therefore has a
molecular weight of approximately $4 \cdot 10^7$ daltons.   The latter value can be

related to the length of the "average" chromosome, which is 6,690 nm. A complete discrete spectrum of the molecular weights of DNA in human chromosomes on the basis of these considerations is given in Figure 5.8.

In isolating DNA from animal organs one similarly encounters discrete spectra of molecular weights: distribution of the molecular weight values of DNA obtained with different methods is not random, although the natural polydisperse character of the material is often distorted by artifacts resulting from the coarse physicochemical procedures. To paraphrase Cairns, the chromosomal DNA molecules may be looked at but not touched. Sadron [109] has published an extensive summary of molecular weights of calf thymus DNA.

It follows from this summary that molecular weights of isolated DNA fall roughly into two categories, namely $M \approx 10^6$ and $M \approx 1.4 \cdot 10^7$, although DNA with a molecular weight of $2.4 \cdot 10^7$ and $3.6 \cdot 10^7$ is also encountered. Isolation of DNA from calf thymus by gentle procedures yields "supermolecules" with a molecular weight of $4 \cdot 10^7$ [110–117].* On chromatographic fractionation on ECTEOLA-cellulose columns [110], calf thymus DNA yields a spectrum with the same discrete character as that reported by Sadron [109].

Thus, the radiation-induced damage to chromosomal DNA in vivo is assessed by analyzing the fraction of DNA isolated from the cell in a rather degraded and homogenized state. In some cases the radiobiochemical analysis provides valuable qualitative and quantitative data even in the absence of definite answers to such questions as the proportion between extracted and total cellular DNA, the exact part of the molecular weight spectrum (or the linkage group) to which the extract belongs, and whether the damage detected is strictly primary or can also be attributed, and to what extent, to postradiation enzymatic action and the physicochemical procedures used for isolation.

It was found that the effects of irradiating DNA in the complex heterogeneous system of the cell are qualitatively the same as those observed on in vitro irradiation of this compound — namely, damage to the nitrogen bases and sugar-phosphate chains of DNA, hydrogen bonds and supramolecular structures, as well as formation of bonds both within and between the DNA chains [110, 111, 118–125].

The available experimental data enable calculation of the radiochemical yield of DNA lesions in vivo [43, 126–130] according to the formula:

$$G = \eta \left. \frac{dC_D}{dD} \right|_{D \to 0} \text{molecules}/100 \, \text{eV},$$

where C_D is the concentration of DNA lesions, itself a function of the dose D, and η is a coefficient which depends on the choice of units for C_D and D. If D is expressed in krad and C_D in micromoles/l, $\eta = 0.965$ [131].

The dose D is measured quite accurately in radiobiological work; accordingly, determination of the concentration of C_D of lesions used in the calculation of G must be as accurate as possible. In the general case

---

* An original method for isolation of euchromatin and heterochromatin DNA with molecular weights of $1 \cdot 10^3$ and $2 \cdot 10^3$ daltons from various animal tissues was described recently (Radiobiologiya, 11 (5). 1971).– The Editor.

$C_D = KC_o$, where C_o is the concentration of DNA-bound nucleotides in the irradiated organ or tissue, and K is the fraction of them damaged by the dose D. Further, $K = \dfrac{m}{M_D} x$, where m is the molecular weight of one nucleotide, M_D the average molecular weight of the region between two lesions of bases or sugar-phosphates, and x is the correction for the loss of DNA during extraction; x equals unity if the DNA is totally extracted, 2 if one half is extracted, etc. The value of C_o varies widely among different tissues of the same organism and in the same tissues of different organisms [132]. Thus, the DNA content of thymus is on average 2.4% in the calf, 2% in the rat and 2% in the rabbit, while the corresponding values for different tissues of the rat are 1.3% in the spleen, 0.23% in the liver, 0.15% in heart muscle, and 0.06% in skeletal muscle. The DNA content of rat hepatoma cells is 2.5 times greater than that of normal liver cells. A DNA concentration of $C_o = 1\%$ by weight corresponds to a concentration of bound nucleotides of $3.3 \cdot 10^4$ micromole/kg.

The radiochemical yield of DNA lesions in vivo has often been compared with the ionic yield of the primary reactions as determined by biophysical analysis of the mutation process at the very beginning of molecular radiobiology [2−5, 133, 134). It became clear that both approaches, namely, the hit and target principles and the physicochemical analysis of the lesions in the isolated DNA, provide the same answer: the ionic (radiochemical) yield of primary reactions responsible for the primary radiobiological process is roughly one molecule per pair of ions or three molecules per 100 eV of absorbed energy. These conclusions, however, are not universally accepted. For this reason a more detailed calculation of the yield of DNA lesions in vivo will be given below.

Radiochemical lesions of bases in cell DNA. The sequence of three-letter "words" out of the four-letter "alphabet" of bases in the chromosomal DNA chain contains genetic information on the amino acid sequence in polypeptides of the highly specific proteins which are essential for normal activity of the cell, organ, tissue and organism. Hence the great interest in the radiation-induced disturbance of this information by the change of even a single "letter" in a single "word" of a long code. Lesions of nitrogen bases in DNA have been obtained by irradiation of plants [110, 111, 118−122], animals [123, 124], and bacteria [125]. These findings indicate that ionizing radiations affect all bases of the cell DNA, though the pyrimidine bases are more susceptible than the purine ones.

A change in the nucleotide composition of the cell DNA of animals can be detected at radiation doses of about 1 krad, and in plants and bacteria at doses exceeding 10 krad. Irradiation of warm-blooded animals causes excretion of decomposition products of the damaged nucleic acids − inosine, xanthine, and uridine [135] − an indication of the modifying effect of radiation on the DNA bases in vivo. Hypoxanthine was found in the DNA of neutron-irradiated pea seeds [121]. Thus, although the chromosomal DNA of irradiated cells can be attacked not only by the free-radical and molecular products of water radiolysis but also by more complex organic free radicals, it appears that radiolytic products of DNA in vivo and in vitro have a similar qualitative composition. The combined efforts of radiation

biochemistry and radiation chemistry should provide a general understanding of the adverse effects of radiation on the DNA bases and genetic
information.

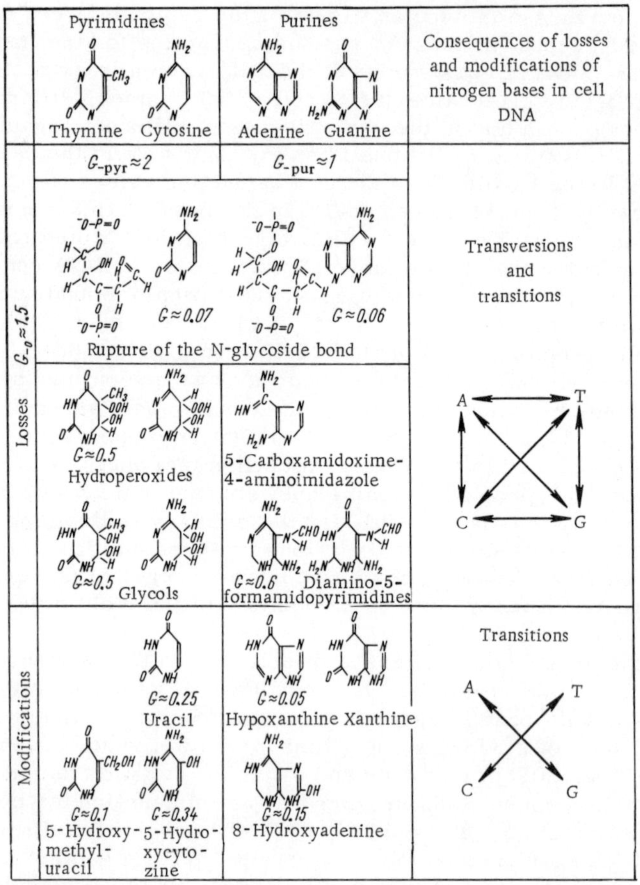

FIGURE 5.9. Radiation-induced damage to DNA bases (losses and chemical
modifications) and its possible consequences (transversions and transitions)

The qualitative effects of irradiation on the DNA bases in vivo and in
vitro are shown in Figure 5.9 by the structural formulae of the radiolytic
products. All radiolytic products may be divided in two groups — losses
and chemical modifications — according to the change in the code triplet.

The losses include not only detachment and loss of bases from the sugar
moiety but also the decomposition products of pyrimidine and purine bases.
This category includes hydroperoxides and glycols of pyrimidine bases,
formamidopyrimidine derivatives of purine bases formed by cleavage of
the imidazole ring (on irradiation in the absence of oxygen), and 5-carboxa-
midoxime-4-aminoimidazoles, formed by cleavage of the pyrimidine ring

of purines (on irradiation in the presence of oxygen). According to the evidence of molecular biology, these compounds lack specific complementarity and can pair with any of the four bases.

The modifications comprise bases containing intact rings but with substitutions in the attached radicals. This group includes the deaminated derivatives of cytosine, adenine and guanine — uracil, hypoxanthine and xanthine, as well as the hydroxyl derivatives of all the bases: 5-hydroxycytosine, 5-hydroxymethyluracil, and 8-hydroxypurines. According to the evidence of molecular biology, some of the modified bases lack the specificity of the initial compounds but retain the general specificity of the bases: they pair not with the partners of the original compounds but with their analogs. For example, hypoxanthine pairs with cytosine, and both hydroxycytosine and uracil pair with adenine; xanthine can pair with thymine.

The yield of certain products, shown in Figure 5.9 according to data obtained from radiolysis of the free bases, provides a rough idea of the probability of particular losses and chemical modifications of the bases in DNA.

Lesions in the bases of the DNA template can bring about changes in the nucleotide composition of de novo synthesized RNA and DNA molecules, thus altering the amino acid composition of newly synthesized proteins not only in the irradiated cells but also in the cell populations which develop from them. Figure 5.9 shows all the conceivable variations of single base substitutions, in the form of transitions and transversions (A, G, C, and T are the initials of the four bases). Transversions are substitutions of pyrimidine bases by purine bases or vice versa (pyr $\rightleftarrows$ pur); they include all the changes along the perimeter of the squares shown in the right-hand part of the diagram. Transitions are substitutions of a pyrimidine base by another pyrimidine base or of a purine base by another purine base; they are represented by the diagonal lines in the square.

Quantitative data on the radiolysis of DNA bases in vivo and in vitro are given in Figure 5.10, which shows the concentration of lesions G_D as a function of the dose D. Nofre [125] has made an interesting quantitative study of the damage to DNA bases in vivo. He examined the damage to pyrimidine bases in the DNA of E. coli u^- on irradiation with gamma quanta of Co^{60} both in vivo and in an aqueous solution of DNA isolated from the cells. In the case of in vivo irradiation the repair systems of the cells were inactivated by the addition of specific inhibitors in order to eliminate the modifying effect of biological factors on the radiochemical process. The irradiated DNA had the same concentration, Co, in the cells and in the solution — namely, 1% ($3.3 \cdot 10^4 \, \mu$moles/kg). Irradiation was carried out in an atmosphere of oxygen (O), air (A) and nitrogen (N). The curves 1 and 2 show the experimental values of C at the corresponding doses D.

The same diagram shows data on the lesions of bases in vitro irradiation of DNA from rat liver and chick erythrocytes (see curves 3 and 4 [137, 138]) as well as calf thymus DNA (curves 5 and 6 [139]). Curve 7 shows the loss of bases due to their detachment from the sugar moiety [137, 138]. Here the values of C_D were determined quite accurately according to the concentration of malondialdehyde, which is generated during irradiation of DNA.

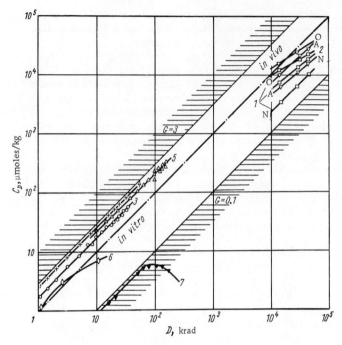

FIGURE 5.10. Effect of the dose of gamma irradiation on the concentration
of lesions in DNA bases

It is evident from the data given in Figure 5.10 that the yields of damage
to DNA bases in vivo and in vitro irradiation are very similar. The ionic
yield of the damage is close to unity, while the radiochemical yield is approximately three bases per 100 eV. As in radiolysis of aqueous solutions,
this yield is evidently limited by the number of free radicals capable of
entering into reaction with the DNA bases. When the concentrations of DNA
in the cell are quite high (about 1–2%), reliable detection of radiation-induced lesions of the bases requires large radiation doses, of the order of
hundreds of kilorads. Endogenous inhibitors and sensitizers apparently
exert only a moderate effect on the yield in the cell.

Radiochemical lesions of the sugar-phosphate chains of cell DNA. Both
in vivo and in vitro, lesions of the sugar-phosphate chains of DNA take the
form of single and double breaks. The breaks can be divided in two categories according to their site — at phosphodiester or carbon-to-carbon
bonds, while as regards the time of their manifestation they can be divided
into actual and potential breaks. Actual single breaks are those induced
directly by radiation at phosphodiester bonds; cleavage of the furanose
ring of deoxyribose with subsequent loss of the base can be assigned tentatively to the category of potential single breaks. This lesion involves
rupture of the $C(3')-C(4')$ bond.

Double breaks result from accumulation of single breaks along the
polynucleotide chains. Actual double breaks can be divided into "straight"
and "oblique" breaks of phosphodiester bonds in the complementary strand

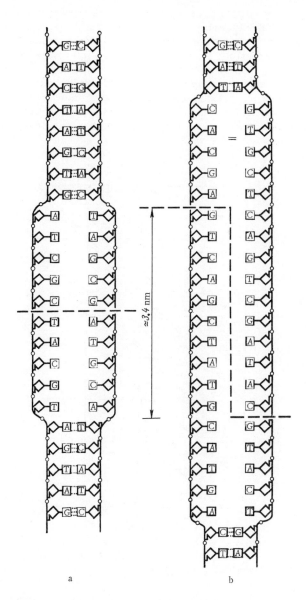

a   b

FIGURE 5.11. Scheme of the straight (a) and oblique (b) breaks
occurring in DNA chains

(Figure 5.11). An oblique break appears immediately if two single breaks in the complementary strands are not more than 10 bases apart. Potential double breaks can arise from a variety of DNA lesions, two of which deserve particular attention: 1) when a latent lesion of deoxyribose appears at a distance of not more than 10 nucleotides from an actual break in the complementary strand; 2) when two actual single breaks or latent lesions in the complementary strands are slightly more than 10 nucleotide pairs apart, but the bases, attached by hydrogen bonds in this area, include the unstable products of their own radiolysis (hydroperoxide-type compounds) which can decompose in the dark.

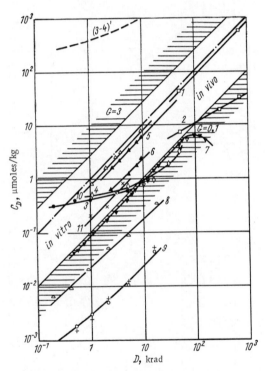

FIGURE 5.12. Effect of the dose of gamma radiation on the concentration of single and double breaks in the sugar-phosphate chains of DNA

Quantitative data on the single and double breaks in DNA polynucleotide chains resulting from irradiation in vivo and in vitro are presented in Figure 5.12 in the form of the relationship between the dose D and the concentration C_D of damaged sugar-phosphate components. The values of C_D were calculated as the product $C_0 \cdot \dfrac{m}{\overline{M}_D} x$ on the basis of data taken from studies in which either $\overline{M}_D$, or $\rho = \dfrac{m}{\overline{M}_D}$, or the number N of single and double breaks per cell was determined. Curves 1–4 refer to lesions of DNA in vivo. After irradiation of leukemic cells of mice in culture (curves 1 and 2), the number N of single and double breaks in the total isolated DNA per cell was determined experimentally [140]. The value of N was converted to $C_D = \dfrac{mC_0}{\bar{M}_0 \, 2n}$ by calculating the factor $\dfrac{mC_0}{\bar{M}_0 \, n}$ at $C_0 = 0.3\%$, the typical value for the blood of warm-blooded animals [136]. The following values were taken: $m = 330$, $\bar{M}_0 = 4 \cdot 10^7$; number of chromonemata $n = 40 \cdot 8 = 320$. The values obtained for $C_D = 2.5 \cdot 10^{-4} \, N$ are plotted against the dose in Figure 5.12. As may be seen, on in vivo irradiation, the yield of single breaks of DNA strands (curve 1) is close to unity, while the yield of double breaks is less than 0.1 (curve 2).

Curves 3 and 4 show a decrease in the viscoelasticity of thymus DNA on irradiation of rats [114, 115, 117]. The following values were taken: $C_0 = 2\%$; $m = 330$, $\bar{M}_0 = 4 \cdot 10^7$. The authors assume that the change in the viscoelasticity of DNA in vivo results from single and double breaks, which lower the molecular weight of the supramolecular structures. However, this finding can also be interpreted in a somewhat different manner by assuming that the supramolecular structure of DNA with an average molecular weight of $4 \cdot 10^7$ represents a steeply coiled, looped chromonema. A double break in the polynucleotide strands possibly impairs the integrity of the loop and destroys some elements of the supercoiling. Such a transition may be accompanied by a decrease in viscoelasticity. The concentration C_D of the nucleotides participating in this mass change can be calculated from curves 3 and 4. This concentration is about 10^3 times greater than the concentration of double breaks if the structural unit for the transition at each break is taken as equal to one gene, i. e., the chromonemal sector enclosed between two chromomeres of the paired spiral helices (see Figure 5.2). In Figure 5.12, such a level of damage is represented by the curve $(3-4)'$ with an average yield $G \simeq 500$ nucleotides per 100 eV.

Curves 5–11 refer to single and double breaks of DNA upon in vitro irradiation of aqueous solutions. The experimental points of curve 5 were obtained by irradiation of aqueous solutions of calf thymus DNA at $C_0 \simeq 0.05\%$ and $M_0 \simeq 5.4 \cdot 10^6$ [141]. The point at the extreme right of this curve was obtained from isolated nuclei of calf thymus [142]. The relative density of single breaks, $\rho = \dfrac{m}{\bar{M}_D}$, has been determined by Ryabchenko et al. and Usakovskaya [141, 142]. In one of these works [142] $\rho = 1.05 \cdot 10^3$ at $D = 40$ krad. If $C_0 = 1\%$ for thymus nuclei, it follows that $C_D = 31.5$ and $G = 0.79$. The points of G on curve 6 were obtained for calf thymus DNA of $M_0 = (8-10) \cdot 10^6$ [143]. For some unknown reason they were below the points of curve 5. Possibly a (three times) higher concentration (about 0.15%) was used in this experiment and this was not taken into account in the calculation of C_D.

Curve 7, like the curve bearing the same number in Figure 5.10, characterizes ruptures of carbon-to-carbon bonds in DNA, which result in loss of bases and liberation of malondialdehyde [138]. Curves 8 and 9 refer to single breaks in phage T2 DNA ($M_0 = 1.2 \cdot 10^8$ and $2.7 \cdot 10^8$) and rat thymus DNA ($M_0 = 2.07 \cdot 10^7$) on irradiation of dilute aqueous solutions at

$C_0 = 0.006\%$ [117]. Curve 10 represents the same parameters but at $C_0 = 0.1\%$ and $M_0 = 7.5 \cdot 10^6$ [144]; finally, curve 11 shows double breaks in calf thymus DNA at $C_0 = 0.1\%$ and $M_0 = 8 \cdot 10^6$ [144].

It is evident from the experimental data shown in Figure 5.12 that single and double breaks in DNA appear with the same probability in the in vivo and in vitro irradiation (curves 1, 5, and 10; 2 and 11). The yield of single breaks in highly dilute DNA solutions (below $C_0 = 0.1\%$) decreases in parallel with the concentration; at $C_0 = 0.006\%$ it amounts to $0.02-0.003$ breaks per 100 eV. In the cell, the yield of single breaks in DNA is slightly less than unity, and that of double and carbon-to-carbon breaks appears to be slightly less than 0.1. The yield of single breaks in protein-free DNA polynucleotide strands is about one third as large as that of damaged bases; the total yield of lesions is approximately 3 nucleotides per 100 eV, with the bases accounting for $^3/_4$ of this amount (2.25), and the sugar-phosphate components for the remaining quarter (0.75). In nucleotides the yield of base lesions is roughly one tenth as high [145], and the corresponding value for the carbohydrate moiety one half as high [143]. Thus, the total yield of lesions of DNA in the nucleoprotein can be estimated at about 0.6 nucleotides per 100 eV of absorbed energy. In nucleoproteins the yield of single breaks of polynucleotide strands (0.375) appears to be 1.5 times that of base lesions (0.225).

Ruptures of hydrogen bonds in cell DNA. The decrease in the hyperchromism and melting temperature of DNA isolated from irradiated cells indicates a radiation-induced denaturation of the biopolymer and a weakening of the hydrogen bonds between its strands. Irreversible ruptures of DNA hydrogen bonds can occur as a result of radiolysis of the bases following breakage of the polymer strands [43, 138]. It has been calculated that each single break in the polymer strand causes the rupture of $13-24$ (on average 19) of the neighboring hydrogen bonds which hold about $8-10$ nucleotide pairs [43, 138, 146]. In view of the equal yield of single breaks of DNA strands in vivo and in vitro it may be expected that the yields of severed hydrogen bonds will also be equal. The yield of severed hydrogen bonds will be about 13 in native, protein-free cell DNA and close to 24 in the partly denatured compound. The corresponding value for protein-bound native DNA can be about $6-7$.

Intra- and intermolecular bonding in cell DNA. Photobiologists have examined the formation of cyclobutane-type thymine dimers in cell DNA exposed to UV radiation [147−152]. Radiobiological studies in this field have failed to clarify the molecular nature and yield of the covalent bonds which appear under the influence of ionizing radiations, although radiation chemistry demonstrates the existence of such bonds. The proof lies in phenomena such as loss of solubility of dry DNA on irradiation in vacuum [153], increase in molecular weight and formation of insoluble gels on irradiation of concentrated solutions of DNA [154]. Electron micrographs of DNA preparations show the formation of branched structures.

Opinions vary as to the nature of the bonds which appear in DNA under the influence of ionizing radiations. Presumably these are covalent bonds which link two reactive ends of the polymer chain and appear at the moment of irradiation, as well as covalent bonds between damaged bases and between

bases and sugar-phosphate residues.   Each of these possibilities is supported by experimental data.

In view of the low yield of single breaks of phosphodiester bonds, the
question arises whether there is a nonenzymatic, strictly radiation-induced
bonding of the ends formed by restoration of phosphodiester bonds.   Broadly
interpreted, this assumption presumes that the nucleotides can polymerize
in any combination under the influence of radiation.   That this is indeed so
can be seen from the formation [155] of oligomers with an average molecular weight roughly equal to that of eight nucleotides following irradiation
of a 0.004% solution of equimolar amounts of the four ribonucleotides
(0.001% of each).   It was found that about 16% of the nucleotides polymerize
by forming a standard phosphodiester bond on irradiation at a dose of
144 krad.   The random combination of sequences of nucleotides in radiation-
induced polymerization perhaps involves not only the standard phosphodiester bonds $C(3')-O-P-O-C(5')$ but also some extraordinary ones, such as
perphosphate bonds of the type $C(3')-O-P-O-O-P-O-C(5')$ between the
normal 5'-phosphate ends and $C(5')-OC(3')-O-$phosphate ends which appear as a result of rupture of the bond caused by radiation [126].   Such processes are known in electrochemistry [156], taking the form $2PO_4^{11}-2e =$
$\rightleftharpoons P_2O_8^{111}$, as well as in radiation chemistry in the dimerization of HSO_4 into
$H_2S_2O_8$ [157].

The existence of radio-induced covalent bonds between damaged bases is
evident from the formation of the thymine radical during photolysis [158]
and radiolysis [159]; this radical can participate in formation of dithymine
peroxides $(C-O-O-C$ bonds between methyl groups), which are possibly the
precursors of thymine dimers [160, 161].   Indeed, the gamma irradiation of
a dry powder of DNA does yield thymine dimers [125].

Various intra- and intermolecular bonds may arise with the participation
of aldehyde groups formed by radiolysis of the carbohydrate moiety of
DNA.   In this case the covalent bonds can unite a deoxyribose residue with
a base or phosphate, as well as two deoxyribose residues.   There are indications that endogenous malondialdehyde (MDA) reacts with the cellular
DNA and alters the structural and functional properties of the latter: MDA
reacts with the guanine and cytosine of DNA and thus prevents its depolymerization by DNase [162]; DNA isolated from cells previously incubated
with MDA is more resistant to DNase and has a higher melting point than
usual [163]; MDA inhibits the replication of DNA, the template synthesis
of RNA, and the growth of cultured human fibroblasts [163–165]; finally,
the MDA formed by irradiation of sucrose can initiate mutagenesis [166] and

shows a cancerostatic effect [167].   The aldehyde group $-\overset{|}{\underset{|}{C}}-CH_2-\overset{O}{\underset{H}{\diagdown}}$

formed directly in the DNA by radiolysis of deoxyribose, can conceivably
be regarded as a more active participant in the formation of intra- and
intermolecular bonds than exoaldehydes; this group can create a covalent
bond with a base or sugar residue in either the same or the complementary
strand.

The quantitative data on radiation-induced bonds are barely more than
estimates.   It was found that the in vitro yield of radiation-induced bonds

in DNA is maximal at a concentration of about 30% [154]. At higher concentrations the free ends cannot meet, whereas at lower concentrations the predominant event is depolymerization caused by products of the radiolysis of water. Oxygen inhibits the formation of bonds, especially in the range of optimal DNA concentrations (30—50%). The yield of bonds was compared with the yield of single and double breaks on irradiation of dried DNA in oxygen, air, and a vacuum [168]. On irradiation in a vacuum, every third event in the single strand gives rise to a bond, while the remaining two thirds cause breaks ($G_{bond} = 0.37$; $G_1 = 0.63$; $G_2 = 0.11$); in oxygen the yield of bonds is only 5% of the yield of breaks ($G_{bond} = 0.16$; $G_1 = 3.4$; $G_2 = 0.16$). In air $G_1 = 1.64$ and $G_2 = 0.13$.

It follows from these data that the yields of bonding and breakage during irradiation of 30—50% DNA in the absence of oxygen are roughly equal. The yield of single breaks in this range of DNA concentrations is comparatively low, about 0.22—0.24 [144], possibly because of the compensating effect of bonding rather than by chance. If the total yield of single breaks and bonds in this range is taken as unity, the yield of bonds alone will be close to 0.75.

Despits its low yield, intra- and intermolecular bonding in cellular DNA can exert a considerable adverse effect on the functions of chromosomes. In protein-bound DNA additional bonds may appear between the DNA and the protein, as can be seen from the difficulties encountered in extracting DNA from irradiated cells.

The yield of DNA lesions in vivo and in vitro. After summing up the data on the lesions of DNA induced by irradiation in vivo and in vitro it is interesting to compare these data on the basis of a single quantitative criterion — namely, the radiochemical yield G, expressed in molecules per 100 eV. The results of such a comparison appear in Figure 5.13, which shows G as a function of the concentration of the irradiated DNA solution. The solid lines represent the radiochemical yield of DNA lesions in vitro within a wide range of concentrations (0.001—100%); the yields of the corresponding DNA lesions in vivo appear as separate regions with the appropriate designations.

It is evident from Figure 5.13 that the radiochemical yield of lesions of the nitrogen bases and sugar-phosphate chains of DNA agrees well with the ionic yield of mutations (dot-dash-dot line) determined by biophysical analysis of the mutation process at the dawn of molecular radiobiology [2—5, 133, 134].

The general pattern of DNA lesions in vivo. The radiochemical lesions of DNA can be outlined as follows on the basis of the literature and our own data on the radiolysis of nucleic acids and their components in vitro, together with the above comparison of DNA lesions in vivo and in vitro.

Irradiation of chromosomal DNA yields at first a complex macroradical with two types of position of the unpaired electrons, corresponding to lesions of the bases — losses and chemical modifications (see Figures 5.9 and 5.10) and lesions of the sugar-phosphate chains — namely, single and double breaks of the phosphoester and intercarbon bonds (see Figures 5.11 and 5.12). With respect to the time of their formation the breaks of the phosphoester bonds can be defined as true lesions which appear directly as a result

of irradiation, whereas the breaks of the intercarbon bonds represent potential lesions manifested as an aftereffect.

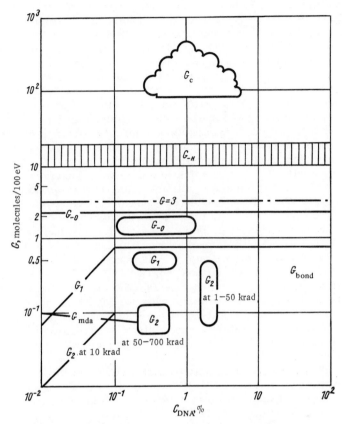

FIGURE 5.13. Comparative data on lesions of DNA in vivo and in vitro:

G_0) lesions of bases; G_1) single breaks of the strands; G_2) double breaks of
the strands; G_{mda}) breaks of intercarbon bonds with the liberation of malondialdehyde; G_H) breaks of hydrogen bonds; G_c) configurational changes;
G_{bond}) bonding.

The primary reactions in the bases and carbohydrate moiety of DNA
are accompanied by secondary reactions including reversible and irreversible breaks of the hydrogen bonds, configurational changes in the
supramolecular DNA structures, and intra- and intermolecular bonding of
the polymer strands.

According to comparable experiments, base lesions are about 3 times
as frequent as lesions of the sugar-phosphate moiety of DNA; the respective yields are 2.25 and 0.75 (see Figure 5.13). Pyrimidine bases are twice
as vulnerable to radiation as purine bases; the respective yields are 3 and
1.5. Radiosensitivity of the phosphodiester bonds is 7.5 times greater than

that of the carbon bonds, the corresponding values being 0.75 and 0.1.   The breakage of hydrogen bonds in protein-free native DNA has a yield of about 18, compared with about 10 in partly denatured DNA; each single true break involves rupture of 13−24 hydrogen bonds between 6−9 base pairs (6 in the GC area, 9 in the AT area).   The yield of double breaks is dose-dependent. On irradiation at a dose of about 10 krad or slightly more this yield amounts to 0.1 of the yield of single breaks.   Double breaks in the circular DNA of the chromonema destroy the supramolecular structure (i. e., the superhelix). The yield of the latter reaction amounts to 100−500 nucleotides per 100 eV (see Figure 5.13).

The reactions leading to losses or modifications of bases and rupture of phosphodiester bonds are not specific to ionizing radiations.   Pyrimidine hydroperoxides and glycols appear also under the influence of certain chemical agents (for example, inorganic peroxides and UV light [147, 148]). Alkylating compounds can break down the pyrimidine and imidazole rings of purine bases [169−173].   Deamination of adenine, guanine and cytosine takes place under the influence of nitrous acid, hydroxylamine and UV irradiation [148, 174].   Phosphodiester bonds are broken even in the normally metabolizing cell as a result of the action of endogenous enzymes (depoly-merases), exogenous chemical agents (alkylating compounds, for example), heat, cold, untrasound and UV irradiation.   The specificity of the inter-carbon breaks is still controversial.   This reaction appears to be highly characteristic of ionizing radiations [43, 127, 130, 138, 139, 146], but there are indications that some chemical agents [169] and UV irradiation [175] can act in a similar manner.   Malondialdehyde, a product of this reaction, can attack DNA and form intra- and intermolecular bonds in it.   This compound appears in the cell as a result of radiolysis of sugars, glutamic acid, and unsaturated fatty acids [176−179].   Accordingly, it can be regarded as a dangerous "radiotoxin" [165].

The relationship between functional **disturbances in the cell and struc-**tural lesions of chromosomal DNA.   Molecular radiobiology regards the structural lesions of chromosomal DNA as responsible not only for gene and chromosome mutagenesis but also for the interphase and reproductive death of cells.

Gene mutations can be attributed to losses and chemical modifications of the bases participating in DNA codons.   Figure 5.14 shows diagram-matically the losses and modifications of bases involved in radiation-in-duced disturbances of the genetic information of the cell.   In fact, this scheme [43, 128, 129] covers all possible transversions and transitions in the DNA codons and therefore exaggerates the actual state of affairs.   Ap-parently, only some of the processes shown in the diagram actually take place, but the available experimental data are insufficient to distinguish between these processes and to establish the nucleotide shifts in the codons and the corresponding changes in the amino acids of the peptides.   The scheme can be used as a program for such experimental work with various models [43, 129].   It is evident at least that irradiation of animals alters the physicochemical properties and amino acid composition of certain proteins [180−188].

Molecular radiobiology advanced deeper into the molecular mechanism of gene mutations with analysis of the amino acid sequences in proteins synthesized on damaged templates. In his 1962 Nobel Speech, Crick [189] noted that his ideas on the genetic code could only be conclusively proved by further study of the mutation-induced changes in the amino acid sequence in the protein molecule. Today, when the code of nucleotides and amino acids has already been deciphered, the above statement can be differently interpreted in the sense that analysis of the amino acid content and sequence of polypeptides is crucial for the understanding of gene mutations.

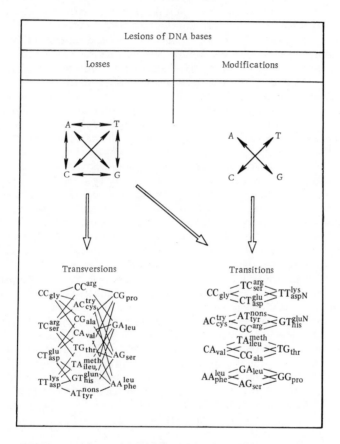

FIGURE 5.14. Scheme of the radiation-induced disturbances of genetic information of the cell associated with losses and chemical modifications of the nitrogen bases of chromosomal DNA

Of particular interest is the analysis of amino acid sequences in polypeptides synthesized after irradiation. Such studies are now under way at L'vov University, where Sukhomlinov and his colleagues have already begun to determine amino acid substitutions in peptides of irradiated animals [190]. The choice of peptide (protein) can be made on the basis of available experimental data. Especially valuable in this respect are rapidly metabolizing proteins such as histones, which participate in the repression-derepression mechanism of genes. Changes in the amino acid composition of tumor cells, in particular those characteristic of leukemia [71], indicate the need for a post-radiation analysis of the amino acid sequence in such cells, as performed by Ingram with spontaneous mutations of human globin genes [191], by Yanofsky [192] with mutations of the tryptophan synthetase locus of the DNA of E. coli, and by Wittman [193] with nitrite mutations of the RNA of tobacco mosaic virus. According to Swingle and Cole [194], losses and chemical modifications of the bases in chromosomal DNA result in discontinuation of the template synthesis of RNA, transcription errors, changes in the nucleotide composition and functions of all types of RNA, changes in the amino acid composition and function of the newly synthesized proteins, and are the main cause of the interphase death of cells.

Chromosomal mutations can be attributed to single or double, actual or potential breaks in the polynucleotide strands of chromosomal DNA and also the concomitant rupture of hydrogen bonds and formation of covalent inter- and intramolecular bonds. These processes can be easily visualized by considering the structural and functional organization of the chromosome, the looped structure of the chromonemal DNA and the progressive coiling taking place during mitosis (see Figure 5.1). These lesions affect in particular the functional accuracy of the recombination mechanisms in dividing cells at the S stage during meiotic and somatic crossovers. Normally the recombination mechanisms operate fairly accurately; the rare gametal and somatic mutations arise spontaneously as a result of purely accidental nonreciprocal (nonmutual and nonidentical) intra- and interchromosomal exchanges of DNA fragments. Irradiation may increase the frequency of nonreciprocal exchanges since the accidental breaks in the polynucleotide strands (or their bonding) may introduce donor-acceptor relationships between the chromosomal subunits participating in the recombinations. This is manifested at the first mitosis after irradiation in the form of subchromatidal, chromatidal and chromosomal aberrations which prevent normal division of the cells and cause their death; the surviving cells show various chromosomal defects such as deletions, duplications, translocations and inversions [4–7, 31, 75, 89, 90, 95, 134].

Further progress in understanding the molecular nature of chromosomal mutations will depend on the success of studies on the structural and functional organization of the chromosome of multicellular organisms and the mechanism of DNA replication. Disturbances of the structure of chromosomes and their subunits may cause not only chromosomal mutations but also the reproductive death of the cells.

MOLECULAR ASPECTS OF RADIATION REPAIR AND PROTECTION

The interrelated phenomena of postradiation repair and antiradiation protection can be treated at two organization levels of life, namely, at the cell level, in terms of restoration of the original structure and functions of radiation-damaged intracellular formations, and at the level of tissues, organs and entire multicellular organisms, the restoration of which comprises not only intracellular repair but also (perhaps mostly) other factors,

such as the proliferation of undamaged cells and their dissemination in the
organism, or a compensatory "cooperation" of different damaged organs
until restoration of their normal functions.   These topics are discussed in
the works of Timofeev-Resovskii and Luchnik [195], Paribok et al. [152, 196],
Korogodin [197], Samoilova [150], Dubrov [151], Strelin and Yarmonenko
[198], and Savich and Shal'nov [199, 201].   Comments on these problems
can also be found in the works of non-Russian authors.   A detailed discussion of the experimental models used for the study of various aspects of
repair is therefore superfluous.   Here we shall consider mainly the molecular aspects of the repair and antiradiation protection of the cell nucleus
and its chromosomes with particular reference to the role of this general
biological mechanism in the antiradiation protection of mammals.   For the
sake of convenience and brevity, restoration at the molecular level (that is,
restoration of the original structure and functions of radiation-damaged
intracellular formations) will be termed repair, and restoration of the
cellular structure of organs and tissues by reproduction and dissemination
of cells (repopulation) — regeneration.

The existence of repair as a general biological phenomenon was discovered by radiobiology and photobiology.   First expressed by Setlow in
1963, the idea of the enzymatic removal of lesions from DNA was soon confirmed by Setlow and Carrier and by Boyce and Howard-Flanders [196].
It was shown that photodimers can be eliminated from DNA either by photo-
reactivation, which requires that the lesion itself or the photoreactivating
enzyme be exposed to shortwave UV light, or by dark repair, which involves
a cut and patch mechanism and does not require light.   The dark repair of
DNA lesions induced by UV light or radiation is a multistage process involving participation of enzymes engaged in the natural metabolism of
nucleic acids.   The coordinated work of these is responsible for the genetically determined resistance of the cell toward radiation and other destructive
factors.   The dark repair system comprises the following enzymes: endo-
nucleases, which recognize and remove the lesion; exonucleases, which
widen the resulting gap; polynucleotide kinases, which phosphorylate the
5'OH end; DNA polymerases, which cause polymerization; and polynucleotide ligases, which join the ends.   These enzymes have been found in
bacteria and mammalian cells.   Their functions are being studied both
directly within cells and in model systems containing damaged DNA and
pure isolated enzymes.   The efficiency and specificity of these enzymes
have been determined both in the cell and in molecular models under different conditions affecting both the DNA and the enzyme.   Of particular
interest in this connection are the attempts to distinguish between reparable
and irreparable lesions of DNA.

Although radiochemical research into the effects of radiation on DNA
in vivo and in vitro is still in progress and may yet reveal further important
and interesting details, some comments on repair of the known lesions of
DNA can already be made at this stage.   To begin with, it appears that
some lesions of DNA bases cannot be repaired at all.   This is the case with
chemical modifications of the bases.   For example, hypoxanthine and uracil
can appear as minor components of DNA as a result of the radiation-
induced deamination of adenine and cytosine.   Irreparable lesions of bases

may lead to gene mutations. On the other hand, some lesions of DNA bases can be repaired in the same way as the pyrimidine UV-dimers. The repair of single breaks in the polynucleotide strands of DNA were first studied a short time ago. As expected [200, 201], ionizing radiations cause both reparable and irreparable breaks in DNA strands. Both single and double breaks can be repaired. This depends on the nature of the exposed ends at the site of cleavage. Rupture of the 3'−O−P phosphoester bond, exposing 3'−OH and 5'− phosphate ends, is apparently the most common and the most easily repaired. Such breaks arise as a result of enzyme action. They are healed without loss of genetic information and therefore the chromosomes and cells retain their ability to replicate and reproduce respectively. The intra- and interchromosomal exchange of DNA fragments which takes place during crossing over results from the appearance of such breaks at particular sites on the chromosome. A different situation prevails as regards breaks of the 5'−O−P bond and those occurring between carbon atoms − a topic discussed in numerous works [43, 126−128, 136−138, 146, 202, 202a]. It is evident from the data obtained that rupture of a 5'−O−P bond yields a 3'-phosphate end, which is rather unnatural in metabolizing DNA. Additional conditions must be satisfied if this lesion is to be repaired without loss of the nucleotide. An attempt to repair radiation-damaged DNA by means of the enzyme polynucleotide ligase was made in Smith's laboratory at Stanford University [203]. It was found that this enzyme can join 3'−OH and 5'−O−P ends at sites of broken phosphoester bonds but not breaks between carbon atoms. In principle, the enzymatic joining of DNA ends formed by rupture of various bonds may encounter three obstacles: 1) a carbonyl group in position 3' in the case of rupture of a 3'−O−P bond; 2) a phosphate group in position 3' in the case of rupture of a 5'−O−P bond; 3) an alkyl phosphate group in the case of a broken intercarbon bond. In the last instance restoration of the structure may involve the loss of one nucleotide from the DNA strand. The operation of ligase may be facilitated by using esterases to eliminate the alkyl phosphate ends at the sites of intercarbon breaks. The radioresistance of cells depends on the ability of their enzyme systems to prepare the damaged DNA for repair and carry out the repair process. Most resistant are bacterial strains possessing a repair system which can heal both single and double breaks in DNA. Repair at the level of the organism depends largely on the survival and multiplication of the damaged cells, which in turn are made possible by the repair of DNA. The success of the antiradiation protection of mammals therefore depends to a large extent on the rational use of chemical and physical factors influencing the repair system of the cells.

Protection of mammals from radiation has its molecular basis in an enzymatic system which is responsible for the dark repair of chromosomal DNA and preserves the unique genetic apparatus of the cells from various adverse factors. Analysis of data on the prophylactic and therapeutic agents used for this purpose shows that their antiradiation effect is usually due to their influence on the mechanism involved in the repair of chromosomal DNA in the cell nucleus. Other mechanisms for protection of the same nuclear structures of the cell are involved in the case of only a small number of chemical agents and are responsible for only a part of the effect obtained.

How do these chemical agents interfere with the repair mechanism on entering the cell? Chemical agents introduced before irradiation must bring about a general physiological shift of the cell toward resistance [204], since reactions of the protector with products of the radiolysis of water (capture of radicals) and with the biological target (removal of unpaired electrons from the radiation-damaged DNA) perform a major role in vitro but are of lesser importance in vivo. This general physiological trans-formation of the cell is manifested in biochemical changes which become evident after introduction of the protector — namely, hypoxia and enhance-ment of the antioxidative activity (AOA) of tissue lipids, slowing down of the synthesis of nucleic acids, a rise in the concentration of sulfhydryl groups in proteins, and elimination of conditions favoring formation of radiotoxins in the cytoplasm and accumulation of carbonyl groups in biopolymers. However, these changes as such hardly perform a crucial role as far as protection from radiation is concerned. In our view, the main factor in the general physiological transformation of the cell toward radioresistance may be mobilization of the repair system of the cell [201]. The most effective radioprotectors such as β-mercaptoethylamine, thiol compounds and cyan-ides are known to cause chromosomal aberrations [204]. This can be inter-preted as direct proof of the intensification of intra- and interchromosomal exchanges and also as an indirect indication of mobilization of the repair system. Organisms irradiated under such conditions can survive because the repair system effectively protects the cells capable of division and re-population and thus brings about regeneration of organs and tissues. Per-haps the best radioprotectors are those causing reciprocal exchanges in DNA prior to irradiation. This hypothesis requires experimental verification.

The therapeutic action of some compounds can be attributed to inter-ference with the repair mechanism either immediately following irradiation or some time afterwards. For example, operation of endonuclease can be intensified by preliminary treatment of DNA facilitating the recognition of hidden lesions by the enzyme. If there are numerous gaps, the operation of exonuclease in widening the gaps in opposite directions along complementary DNA strands can bring about lethal double breaks. This can be avoided by enhancing the action of polynucleotide kinase, which transfers phosphate from ATP to the end of the depolymerized DNA molecule and thus stops the operation of exonuclease. Such an effect can be obtained by introducing ATP immediately after irradiation.

The action of DNA polymerase can be stimulated by the introduction of deoxyribonucleotides in the form of a DNA hydrolysate. Elimination of the alkyl phosphate ends by esterases facilitates the operation of ligase. Finally, exogenous DNA can be used to heal lesions in chromosomal DNA in view of the occurrence of "recombination repair" during certain stages of the cell cycle [196]. The available data indicate that iso-, homo-, and even heterologous DNA entering the cell nucleus can participate in recom-binations with the chromosomal DNA. For this purpose the molecular weight of the exogenous DNA must be at least 10^6 daltons. Chromosomal DNA incorporates only a small fragment of this DNA, which nevertheless serves as a kind of nuclear patching material [43, 201]. It appears that exogenous DNA is incorporated into the genome by the same repair enzymes which carry out intra- and intermolecular exchanges of DNA; hence the requirement for a steric compatibility and complementarity. Such events can be interpreted in a way as transplantations at the molecular level.

Besides the dark repair of DNA, which enables some of the vitally necessary cells to reproduce, restoration at the level of the organism includes such processes as dispersion and settling of surviving bone marrow stem cells all over the irradiated organism, acceleration of their division, and dedifferentiation of cells. The dispersion and resettling of stem cells is particularly important in the case of unevenly irradiated organisms. The shielding of some parts of the body may save the organism by preserving some vital cells and maintaining their capacity for reproduction and repopulation. The local shielding of hematopoietic foci with lead in mammals is especially effective [198]. The nonirradiated parts of the hematopoietic tissue can be regarded as foci of the autotransplantation of bone marrow cells by means of resettling stem cells in the irradiated organism. An artificial increase in the migration capacity of the surviving stem cells may be beneficial. A positive therapeutic effect may also be obtained by the transplantation of foreign bone marrow. The transplantation of a pure fraction of bone marrow stem cells appears especially promising. The cells of this fraction do not yet bear the program of the donor organism. Once introduced into the blood stream of the irradiated organism, they settle and reproduce there, adopting any differentiation program that exists in the particular organ or tissue and thus helping the irradiated organism to survive. Such cells practically do not cause the secondary syndrome known as "transplant vs. host." Division of the surviving stem cells can be accelerated by special stimulatory agents such as leucocytin (glycoalkaloid solanine), proposed by Belousov and tested by Gorizontov [205]. Dedifferentiation of cells consists in their transition from early specialization to a state of division. Although the molecular mechanism is still unknown, this transition undoubtedly involves hormones. The study of cell redifferentiation under the influence of hormones may clear the way for a hormonal therapy of radiation disease.

CONCLUSION

The available data on the structural and functional organization of chromosomes and the lesions of their DNA in vivo and in vitro irradiation indicate that radiochemical modifications and losses of bases and breaks in the sugar-phosphate chains can trigger gene, chromosome and genome mutations and even kill the cell. The interphase death of cells can be attributed to an unrestrained production of abnormal protein by a damaged gene, whereas reproductive death results from chromosomal aberrations which impede mitosis and meiosis.

Knowledge of the molecular basis of the mutational and lethal lesions of cells may reveal the molecular mechanism of restoration and antiradiation protection. Restoration consists essentially in the repopulation of cells at the level of the organism and the repair of DNA lesions at the level of the cell.   The chemical requirements for the repair of radiation-damaged DNA cannot be determined without previous knowledge of the exact nature of the nitrogen base lesions, single breaks in the polynucleotide strands, and other radiochemical transformations of DNA.   Hence the practical value of research into the radiolysis of DNA and its components.

Apart from the enzymatic repair of DNA, the control of radiation damage and protection from radiation include such practically important processes as resettling of surviving bone marrow stem cells in the irradiated organism, acceleration of their division, and dedifferentiation of cells.   Studies on the molecular biology of the gene have produced a vast amount of data on the enzymes engaged in the dark repair of DNA;   molecular developmental biology has provided information on the mechanism of cell differentiation and dedifferentiation;   and finally, molecular radiobiology has established that DNA is the main substrate of radiation damage and repair at the cell level.   These findings indicate new paths in the search for means of anti-radiation protection in the four directions corresponding to the four major factors of cell repopulation.   Further research in these fields requires thorough knowledge of the molecular basis of radiation disease and the interaction of exogenous factors with the cell repair system.

The efficiency of radioprotectors and therapeutic agents must be assessed not only by the survival rate but also according to the frequency of long-range effects in the survivors.   A marked increase in the frequency of somatic mutations can be expected among irradiated individuals which have survived as a result of any radioprotectors or therapeutic agents, including DNA.   No prophylactic or curative agent, whether existing or yet to be discovered by future research, can be expected to prevent or heal the radiation damage by itself.   This task falls into the realm of the enzymatic system of DNA repair;   the best we can do is to understand the function of this cybernetic system perfected by evolution well enough to provide it with modest assistance.

BIBLIOGRAPHY

1. Timofeev-Resovskii, N.V.— Tsitologiya 2 (1960), 45.
2. Timofeef-Ressovsky, N.W.— Nova Acta Leopoldium, 9 (1940), 60.
3. Timofeeff-Ressovsky, N.W., K.G.Zimmer, and D.Delbruck.— Nachr. Akad. Wiss. Göttingen, VI N.F. 1 (1935), 189.
4. Timofeeff-Russovsky, N.W. and K.G.Zimmer.   Das Trefferprinzip in der Biologie, S. Leipzig, 1947.
5. Lea, D.E.   Action of Radiations on Living Cells.   New York.   Cambridge Univ. Press. 1962.
6. Prokof'eva-Bel'govskaya, A.A.— In: "Aktual'nye voprosy sovremennoi genetiki," p.23.   Moscow, MGU, 1966.
7. Prokof'eva-Bel'govskaya, A.A.   Principles of Human Cytogenics.—Moscow, "Meditsina," 1969. (Russian)
8. Ris, H.— In: Chemical Basis of Heredity.   Edited by T.L.Knunyanets and B.N.Sidorov, p.25.   Moscow, I.L., 1960. (Russian translation)
9. Taylor, D.C.— In: "Molecular Genetics," p.78.   Edited by A.N.Belozerskii, Part 1.   Moscow, "Mir," 1964. (Russian translation)
10. Levitskii, G.A.— In: "Klassiki sovetskoi genetiki 1920—1940."   Edited by P.I.Zhukovskii, p.171. Moscow, "Nauka," 1968.

11. Navashin,S.G. Symposium in Honor of K.A.Timiryazev,1916. (Cited from [10]). (Russian)
12. Kol'tsov,N.K. Organization of the Cell.— Moscow—Leningrad, Gosmedizdat,1936. (Russian)
13. Prokof'eva-Bel'govskaya,A.A.— In: "Itogi nauki," No.3, p.7. Moscow, AN SSSR,1960.
14. Caspersson,T. Cell Growth and Cell Function.— New York—Norton, 1950.
15. Caspersson,T.— Cold Spring Harb. Symp. Quant. Biol. 21 (1956),23.
16. Demerec,M.— Heredity 24 (1933),33.
17. Darlington,C.D.— Nature 149 (1942),66.
18. Avery,O.T., M.C.McLeod,and M.McCarty.— J. Exp. Med. 79 (1944),137.
19. Watson,J.D. and F.H.C.Crick.— Nature 171 (1953),737,964.
20. Wilkins,M.H.F., A.R.Stokes,and H.R.Wilson.— Nature 171 (1953),738.
21. Steffensen,D.M.— Int. Rev. Cytol. 12 (1961),162.
22. Spirin,A.S.— In: "Uspekhi biologicheskoi khimii," Vol.4,p.93. Moscow, AN SSSR,1962.
23. Bogdanov,Yu.F., A.V.Iordanskii,and V.M.Gindilis.— Genetika, No.5 (1965),81.
24. Gurskii,G.V. and A.G.Malenkov.— Genetika, No.7 (1966),134.
25. Lwoff,A.— J. Gen. Microbiol. 17 (1957),339.
26. Jacob,F. et al.— Cold Spring Harb. Symp. Quant. Biol. 28 (1963),329.
27. Braun,W. Bacterial Genetics. 2nd edition. Philadelphia. Saunders. 1965.
28. Cairns,J. In: Coult,D.A.— Molecules and Cells. London,Longmans,1966.
29. Watson,J.O. Molecular Biology of the Gene. New York. Benjamin 63, 1965.
30. Kornberg,A.— In: "Molecules and Cells." Edited by G.M.Frank,No.4,p.70.— Moscow, "Mir," 1969.
        (Russian translation)
31. Sidorov,B.N. and N.N.Sokolov.— Radiobiologiya 3 (1963),415.
32. De, D.N. — Proc. XII Intern. Congr. Genetics, Tokyo 2 (1968),72.
33. Taylor,J.H.— Ibid., p.25.
34. Barigozzi,C.— Ibid.,p.71.
35. Cummons,J.E.— Ibid.,p.68.
36. Tanaka,N. and A.Kusanagi.— Ibid.,p.69.
37. Pelling,C.— Ibid.,p.74.
38. Bresler,C.E. Introduction to Molecular Biology.— New York. Academic Press 1971.
39. Vol'kenshtein,M.V. Molecules and Life.— Moscow, "Nauka," 1965. (Russian)
40. Medvedev,Zh.A. Mechanisms of Molecular Genetic Development.— Moscow, "Meditsina," 1968.
        (Russian)
41. Mosolov,A.N.— Genetika 4 (1968),135.
42. Polyakov,Yu.V.— Thesis. Moscow,1969.
43. Shal'nov,M.I.— Thesis. Moscow,1969.
44. Young,E.T. and R.L.Sinsheimer.— J. Molec. Biol. 33 (1968),49.
45. Thomas,K.A.— In: Molecular Genetics. Edited by A.N.Belozerskii,Part 1,p.126. Moscow, "Mir," 1964.
        (Russian translation)
46. Tikhonenko,T.I. Biochemistry of Viruses.— Moscow, "Meditsina," 1966. (Russian)
47. Tikhonenko,A.S. Ultrastructure of Viruses and Bacteria.- Moscow, "Nauka," 1968. (Russian)
48. Bonner,J.F. The Molecular Biology of Development. New York. Oxford. 1965.
49. De Rosnay,J.— Atomes 24 (1969),315.
50. Wolstenholme,D.R., I.Dawid,and H.Ristow. — Genetics 60 (1968),759.
51. Kol'tsov,N.K.— Biol. Zh. 7,1937.
52. Kol'tsov,N.K.— Biol. Zh. 7,1938.
53. Miller,O.L. and B.Biti.— Nauka i Zhizn', No.10 (1969),58.
54. Jacob,F. and J.Monod.— Proceedings of the 5th International Biochemical Congress Sympos. 1.,
        Moscow,1962.
55. Butler,J.— In: "Histones." Edited by Engel'gard,V.A.,p.11. Moscow, "Mir," 1968. (Russian translation)
56. Bonner,D. and R.Huang.— Ibid.,p.28.
57. Griffith,A. and J.Bonner. Cited from [49].
58. Zubay,G. The Nucleohistones. Editors J.Bonner and P.Ts'o.— San Francisco,Holden-Day,1964.
59. Davies,H.G. and J.V.Small.— Nature 217 (1968),1122.
60. Littau,V.C. et al.— Proc. Natn. Acad. Sci. U.S.A. 54 (1965),1204.
61. Yuodka,B.A. et al.— Biokhimiya 33 (1968),907.
62. Taylor,J.H. Cytogenet. Cells Cult.,p.175.— N.Y.—L., Academic Press,1964.

63. Bonner, J. and A.E.Mirsky. The Nucleohistones. Editors J.Bonner and P.Ts'o, p.72.— San Francisco, Holden-Day, 1964.
64. Swift, M. The Nucleohistones. Editors J.Bonner and P.Ts'o, p.54.— San Francisco, Holden-Day, 1964.
65. Nilhans, W.G. and C.P.Barnum.— Exptl. Cell Res. 39 (1965), 435.
66. Baer, D.— J. Theoret. Biol. 6 (1964), 282.
67. Evans, J.H. and S.Rogers.— Exptl. and Molec. Pathol. 7 (1967), 105.
68. Raaf, J. and J.Bonner.— Arch. Biochem. Biophys. 125 (1968), 567.
69. Hancock, R.— Bull. schweiz. Akad. med. Wiss. 24 (1969), 386.
70. Butler, J.A. and E.W.Johns.— Biochem. J. 91 (1964), C15.
71. Sahasrabudhe, M.B.— J. Scient. Ind. Res. 26 (1967), 299.
72. Irvin, J. et al.— Explo. Cell Res. 9 (1963), 359.
73. Killander, D. and A.Letterberg.— Exptl. Cell Res. 38 (1965), 272.
74. Levintal, C. and H.R.Crane.— Proc. Natn. Acad. Sci. U.S.A. 42 (1956), 436.
75. Gershkovich, I.— Genetika. Moscow, "Nauka." 1968.
76. Gamov, G.— Proc. Natn. Acad. Sci. U.S.A. 41 (1955), 7.
77. Bloch, D.P.— Ibid., p.1058.
78. Schrödinger, A. What is Life and Mind and Matter. Cambridge. Univ. Press. 1967.
79. Delbrueck, M. and G.Stent.— In: "Chemical Basis of Heredity." Edited by I.L.Knunyanets and B.N.Sidorov, p.562. Moscow, I.L., 1960. (Russian translation)
80. Luchnik, N.V., Yu.M.Plishkin, and G.G.Taluts.— Tsitologiya 2 (1960), 57.
81. Plishkin, Yu.M., N.V.Luchnik, and G.G.Taluts.— Biofizika 6 (1961), 257.
82. Timofeev-Resovskii, N.V. and R.R.Rompe.— Problemy Kibern., No.2 (1959), 213.
83. Orlov, A.N. and S.N.Fishman.— Tsitologiya 2 (1960), 68.
84. Kobelev, L.Ya. and Yu.M.Udachin.— Biofizika 9 (1964), 649.
85. Vol'kenshtein, M.V. and A.M.El'yashevich.— Biofizika 6 (1961), 513.
86. Lima-de Faria, A.— Hereditas 47 (1961), 674.
87. Whitten, J.M.— Nature (L) 208 (1965), 1019.
88. Gilbert, W. and D.Dressler.— Cold Spring Harb. Symp. Quant. Biol. 32 (1968), 473.
89. Sidorov, B.N. and N.N.Sokolov.— Radiobiologiya 4 (1964), 828.
90. Nemtseva, L.S.— Radiobiologiya 5 (1965), 126.
91. Low, B.W.— In: "Chemical Basis of Heredity." Edited by I.L.Knunyanets and B.N.Sidorov, p.458. Moscow, I.L., 1960. (Russian translation)
92. Stonehill, E.H.— J. theoret. Biol. 9 (1965), 323.
93. Zyryanov, P.S.— Tsitologiya 2 (1960), 62.
94. Zyryanov, P.S.— Biofizika 6 (1961), 495.
95. Lobashov, M.E. Genetics.— Leningrad, LGU, 1967. (Russian)
96. Dubinin, N.P. Problems of Radiation Genetics.— Moscow, Atomizdat, 1961. (Russian)
97. Jehle, H.— Proc. Natn. Acad. Sci. U.S.A. 50 (1963), 516.
98. Chentsov, Yu.S. and V.Yu.Polyakov.— Doklady AN SSSR 189 (1969), 185.
99. Unuma, Tadio.— J. Electron Microsc. 18 (1969), 25.
100. Henriks, D.N. and D.T.Mayer.— Proc. Soc. Exp. Biol. Med. 119 (1965), 769.
101. Collins, R.L.— J. Hered. 60 (1969), 117.
102. Zachan, H.G.— Muenchen. Med. Wochenschr. 111(29)1513. 1969.
103. Griffith, J. and J.Bonner. Cited from [49].
104. Morgan, T.H. Mechanism of Heredity. Proc. Roy. Soc. (London) 946, 162—97 (1922). GIZ, 1923. (Russian translation)
105. Morgan, T.H. The Theory of Evolution in Current Interpretation. Edited by S.S.Chetverikov.— Leningrad, GIZ, 1926. (Russian translation)
106. Morgan, T.H. Theory of the Gene. Yale Univ. Press. 1976.
107. Giles, N.H.— In: "The Genetic Apparatus of the Cell and Some Aspects of Ontogenesis." Edited by D.K.Belyaev and V.V.Khvostova, p.5. Moscow, "Nauka." 1968. (Russian translation)
108. Wu, R. and A.Kaiser.— J. Molec. Biol. 36 (1968), 523.
109. Sadron, Charles.— In: "Nucleinic Acids." Edited by N.N.Belozerskii, p.7. Moscow, I.L., 1962. (Russian translation)
110. Kuzin, A.M. Radiation Biochemistry.— Moscow, AN SSSR, 1962. (Russian)
111. Kuzin, A.M.— In: "Osnovy radiatsionnoi biologii," p.51. Moscow, "Nauka." 1964.

112. Kuzin,A.M., N.B.Strazhevskaya,and V.A.Struchkova.— Radiobiologiya 1 (1961),10.
113. Strazhevskaya, N.B.and V.A.Struchkov.— Radiobiologiya 2 (1962),9.
114. Strazhevskaya, N.B., V.A.Struchkov,and G.S.Kalendo.— Radiobiologiya 6 (1966),783.
115. Struchkov,V.A.— In: "Nukleinovye kisloty i biologicheskoe deistvie ioniziruyushchei radiatsii," p.56. Moscow, "Nauka." 1967.
116. Struchkov,V.A. and N.B.Strazhevskaya.— Radiobiologiya 1 (1961),172.
117. Struchkov,V.A. and N.B.Strazhevskaya.— Radiobiologiya 7 (1967),819.
118. Tokarskaya,V.I.— Radiobiologiya 1 (1961),193.
119. Tokarskaya,V.I.— Ibid.,p.330.
120. Tokarskaya,V.I.— Radiobiologiya 2 (1962),161.
121. Tokarskaya,V.I.— Radiobiologiya 5 (1965),566.
122. Tokarskaya,V.I. and A.M.Kuzin.— Radiobiologiya 6 (1966),3.
123. Kritskii,G.A.— In: "Nukleinovye kisloty i biologicheskoe deistvie ioniziruyushchei radiatsii," p.44. Moscow, "Nauka." 1967.
124. Komar,V.E.— Radiobiologiya 3 (1963),816.
125. Nofre,C. La radiolyse de la thymine et ses implications radiophysiologiques.— La faculté des sciences de l'université de Lyon,p.182,1968.
126. Shal'nov,M.I.— Trudy Mosk. Obshch. Ispyt. Prir. 7 (1963),47.
127. Shal'nov,M.I.— In: Pervichnye radiobiologicheskie protsessy,p.52. Moscow, Atomizdat,1964.
128. Shal'nov,M.I.— In: Nukleinovye kisloty i biologicheskoe deistvie ioniziruyushchei radiatsii,p.28. Moscow, "Nauka." 1967.
129. Shal'nov,M.I. and A.V.Savich.— Radiobiologiya 7 (1967),698.
130. Shal'nov,M.I. and N.P.Krushinskaya.— Izv. AN SSSR,ser. biomed. sciences, No.5, Issue 1 (1967),109.
131. Savich,A.V.— In:"Pervichnye radiobiologicheskie protsessy," p.7. Moscow, Atomizdat,1964.
132. Davidson,J.N. Biochemistry of Nucleic Acids. 5th ed. New York, Wiley. 1965.
133. Zimmer,K.G. Studies on Quantitative Radiation Biology. Translated by H.D.Griffith. London, Oliver and Boyd. 1961.
134. Timofeev-Resovskii,N.V., V.I.Ivanov,and V.I.Korogodin. Application of the Hit Principle in Radiobiology.— Moscow, Atomizdat,1968. (Russian)
135. Rysina,T.N.— In: "Biologicheskoe deistvie radiatsii i voprosy raspredeleniya radioaktivnykh izotopov," p.9. Moscow, Gosatomizdat,1961.
136. Krushinskaya,N.P.— Thesis. Moscow,1969.
137. Krushinskaya,N.P. and M.I.Shal'nov.— Radiobiologiya 7 (1967),24.
138. Krushinskaya,N.P. and M.I.Shal'nov.— Radiobiologiya 9 (1969),339.
139. Collins,B. et al.— Radiat. Res. 25 (1965),526.
140. Veatch,W. and S.Okada.— Biophys. J. 9 (1969),330.
141. Ryabchenko,N.I. and P.I.Tseitlin.— Radiobiologiya 3 (1963),153.
142. Usakovskaya,T.S.— Radiobiologiya 5 (1965),791.
143. Usakovskaya,T.M. et al.— Radiobiologiya 6 (1966),489.
144. Hagen,U.— Strahlentherapie 124 (1964),428.
145. Stacey,K.A.— In: "Radiation Effects in Physics, Chemistry and Biology." Edited by P.D.Gorizontov, p.114. Moscow, Atomizdat,1965. (Russian translation)
146. Krushinskaya,N.P. and M.I.Shal'nov.— In: "Deistvie ioniziruyushchei radiatsii na belki i nulkleinovye kisloty. Molekulyarnye mekhanizmy zashchity." Kiev, "Naukova Dumka." 1970.
147. Setlow,R.B. and E.Pollard. Molecular Biophysics. Reading, Mass. Addison Wesley Pub. Co. 1967.
148. Bakker,A.— In: "Molecular Mechanisms of the Effect of Radiation Nucleic Acids." Edited by A.N.Belozerskii,Vol.1,p.142. Moscow, "Mir," 1965. (Russian translation)
149. Zavil'gel'skii,G.B.— In: "Molekulyarnaya biofizika," p.137. Moscow, "Nauka." 1965.
150. Samoilova,E.A. The Effect of Ultraviolet Radiation on the Cell.— Leningrad, "Nauka." 1967. (Russian)
151. Dubrov,A.P. Genetic and Physiological Effects of Ultraviolet Radiation on Higher Plants.— Moscow, "Nauka." 1968. (Russian)
152. Zhestyanikov,V.D. The Restoration and Radioresistance of the Cell.— Leningrad, "Nauka." 1968. (Russian)
153. Setlow,R. and B.Doyle.— Biochim. biophys. Acta 15 (1954),117.
154. Baeq,Z. and P.Alexander. Fundamentals of Radiobiology. 2nd ed. New York, Pergamon Press. 1961.

155. Contrera, G. et al.— Biochim. biophys. Acta 61 (1962), 718.
156. Remy, H. Lehrbuch der anorganischen Chemie.— Leipzig, 1957.
157. Daniels, M. et al. Radiation.— J. Chem. Soc., (1957), 4388.
158. Daniels, M. and A.Grimison.— Biochem. Photobiol. 5 (1964), 119.
159. Latarjet, R. et al.— Radiat. Res. 3 (1963), 247.
160. Wang, S.Y. and P.Alcantara.— Photochem. Photobiol. 4 (1965), 477.
161. Wang, S.Y. et al.— Proc. Natn. Acad. Sci. U.S.A. 57 (1967), 465.
162. Brooks, B.B. and O.Z.Klamerth.— Europ. J. Biochem, 5 (1968), 178.
163. Klamerth, O.Z. Quantitativ. Biol. Metabol., p.177.— Berlin, Heidelberg—New York, 1968.
164. Klamerth, O.Z. and H.Levinsky.— Biochim. biophys. Acta 155 (1968), 271.
165. Klamerth, O.Z. and H.Levinsky.— FEBS Letters 3 (1969), 205.
166. Shaw, M.W. and E.Hayes.— Nature 211 (1966), 1254.
167. French, F.A. and B.L.Freedlander.— Cancer Res. 18 (1958), 172.
168. Hagen, U. and H.Wellstein.— Strahlentherapie 128 (1965), 565.
169. Bartoshevich, Yu.E.— In: "Supermutagens," p.211. Moscow, "Nauka." 1966.
170. Gumanov, L.L.— In: "Uspekhi sovremennoi genetiki," No.2, p.96. Moscow, "Nauka." 1969.
171. Fahmy, O.G. and M.J.Fahmy.— Ann. N.Y. Acad. Sci. 160 (1969), 228.
172. Loveless, A.— Nature 223 (1969), 206.
173. Süssmuth, R. and F.Lingens.— Z. Naturf. 24b (1969), 903.
174. Fritz, E.— In: "Molecular Genetics." Edited by A.N.Belozerskii. Part 1, p.226. Moscow, "Mir." 1964.
     (Russian translation)
175. Täufel, K. and K.Zimmermann.— Zeitschrift Lebensmittel-Untersuch. und Forsch. 118 (1962), 8.
176. Ambe, K. and A.Tuppel.— Fd. Sci. 26 (1961), 448.
177. Morre, J. and Morrazotti-Pelletiers.— C.r. hebd. Séanc. Acad. Sci., Paris, 262 (1966), 1729.
178. Philpot, J. St. Z.— Radiat. Res. 3 (1963), 55.
179. Millos, N.A., R.S.Harris, and A.Golubovic.— Radiat. Res. 3 (1963), 71.
180. Blokhina, V.D. and M.I.Shal'nov.— Byull. Eksp. Biol. Med. 48 (1959), 49.
181. Waldschmidt-Leitz, E. and L.Keller.— Strahlentherapie 116 (1961), 610.
182. Sukhomlinov, B.F., V.R.Merenov, and N.N.Semenchuk.— In: "Biologicheskoe deistvie radiatsii,"
     No.1, p.3. Lvov University. 1962.
183. Sukhomlinov, B.F. and A.N.Yakovenko.— Ibid., No.2, p.3, 1963.
184. Sukhomlinov, B.F., Ya.V.Oleinik, and M.P.Pokosh.— Ibid., p.26.
185. Datskiv, M.Z.— Ibid., No.3, p.21, 1965.
186. Makovetskii, N.I.— Ibid., No.4, p.46, 1969.
187. Petrova, N.D. and L.T.Tutochkina.— In: "Proceedings of the 8th Intern. Cancer Congress," Vol.4, p.13.
     Moscow—Leningrad, Medgiz, 1963. (Russian)
188. Olontseva, O.I.— Radiobiologiya 11 (1971), 499.
189. Crick, F. The Genetic Code. Edited by N.I.Shapiro. Suppl. to "Genetics" by I.Gorshkovich, p.653.—
     Moscow, "Nauka." 1968. (Russian translation)
190. Sukhomlinov, B.F.— Inform. Byull. "Radiobiologiya," No.13, 1970. (In press)
191. Ingram, Vernon M. The Biosynthesis of Macromolecules. New York, Benjamin. 1965.
192. Yanofsky, C.— In: "Information Macromolecules." Edited by V.A.Engel'gard, p.154. Moscow, "Mir."
     1965. (Russian translation)
193. Wittmann, H.— In: "Information Macromolecules," p.140. Edited by V.A.Engel'gard. Moscow,
     "Mir." 1965. (Russian translation)
194. Swingl, K.E. and L.I.Cole. Current Topics in Radiation Research, Vol.4, p.184.— Amsterdam, North-
     Holland Publishing Company, 1968.
195. Timofeev-Resovskii, N.V. and N.V.Luchnik.— In: "Trudy Inst. biologii UFAN SSSR," Vol.9,
     p.57. Sverdlovsk.
196. Kuzin, A.M. (Editor). Current Problems in Radiobiology, Vol.1. Postradiation Repair. Edited by
     V.P. Paribok.— Moscow, Atomizdat, 1970. (Russian)
197. Korogodin, V.I. Problems of Postradiation Repair.— Moscow, Atomizdat, 1966. (Russian)
198. Strelin, G.S. and S.P.Yarmonenko.— In: "Sovremennye problemy radiobiologii." Edited by
     A.M.Kuzin, Vol.1. Postradiatsionnaya reparatsiya, p.264. Edited by V.P.Paribok. Moscow,
     Atomizdat, 1970.
199. Savich, A.V.— Inform. Byull. "Radiobiologiya," No.12 (1969), 3.

200. S a v i c h , A.V. and M.I.S h a l ' n o v.— In: "Vosstanovitel'nye protsessy pri porazhenii organizma
     ioniziruyushchei radiatsiei," p.33. Moscow, 1966.
201. S h a l ' n o v , M.I.— In: "Otdalennye posledstviya luchevykh porazhenii." Edited by Yu.I.Moskalev, p.37.
     Moscow, Atomizdat, 1971.
202. K a p p , D.S. and K.C.S m i t h.— Radiat. Res. 42 (1970), 34.
202a. U l l r i c h , H. and U.H a g e n.— Int. J. Radiat. Biol. 19 (1971), 507.
203. K a p p , D.S. and K.C.S m i t h.— Int. J. Radiat. Biol. 14 (1969), 567.
204. B a c q , Z.M. Chemical Protection against Ionizing Radiation.— Thomas, 1965.
205. G o r i z o n t o v , P.D. et al.— Patol. Fiziol. Eksperim. Terap. 15 (1971), 54.

Chapter 6

RADIATION DAMAGE TO CELL MEMBRANES

INTRODUCTION

Membranes perform an important role in the structural and functional organization of the cell. Hence the great interest in the experimental assessment of radiation-induced damage to membranes within the general framework of radiobiological effects. Radiation-induced damage to cell membranes is regarded today as one of the most important primary mechanisms of the effect of radiation on the cell.

Intensive research into the nature of biological membranes has confirmed that the functioning of the living cell depends on its structural organization, which in turn is determined by the heterogeneous, multiphase internal structure of the protoplasm. Apart from completing the structural organization of the cell, membranes ensure the strict sequence of metabolic events enabling the cell to regulate its activity and maintain its structure and functions in a changing environment.

New methods and approaches in research have enabled the isolation of various biological membranes in a pure state, thus contributing greatly toward a better knowledge of the chemical composition, biochemical properties and functions of cell membranes. The principal membranes of the cell are the surface, or plasma membrane, the endoplasmic reticulum, the mitochondrial membranes, the nuclear membrane, the lysosomal membrane, the Golgi apparatus, etc. The membranes of animal cells are shown diagrammatically in Figure 6.1 [1].

All cell membranes appear to be constructed according to a common pattern with a characteristic triple-layer lipoprotein structure as illustrated in the "unit" membrane proposed by Robertson. Electron microscopy has shown the membrane to be composed of two layers of oriented lipid molecules flanked on either side by protein monolayers. However, the actual structure of the membrane is more complex [1—4]. An increasing body of evidence indicates that biological membranes perform a multitude of functions and are directly or indirectly related to all the processes occurring in the living cell [5—7].*

---

* Data obtained by the combined use of electron microscopy and autoradiography [Comings, D.E. Amer. J. Hum. Genet. 20:440—60. 1968; Mosolov, A.N., Genetika 4:135. 1968] have led to the hypothesis that interphase chromosomes are attached to the nuclear membrane. Convincing proof of this hypothesis was furnished by Japanese workers [Kitayama S. and A.Matsuyama, Int. J. Rad. Biol., 19:13. 1971], who isolated the membrane fraction of DNA (90% of the total cell DNA) from Micrococcus radiodurans by means of Triton X-100. Further support was provided by the work of Ormerod and Lehman (Biochim. Biophys. Acta, 228:331. 1971), who found that high molecular weight DNA of L5178Y animal cells is attached to the nuclear membrane.— The Editor.

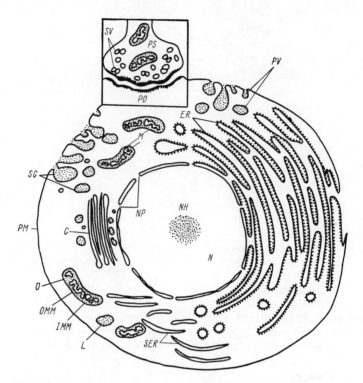

FIGURE 6.1. Scheme of the animal cell membranes normally visible under the electron microscope [1]:

C) mitochondrial cristae; ER) rough endoplasmic reticulum; G) Golgi apparatus; IMM) inner mitochondrial membrane; L) lysosome; M) mitochondrion; N) nucleus; NP) nuclear pores; NH) nucleolus; OMM) outer mitochondrial membrane; PM) plasma membrane (cell membrane, or surface membrane, or cytoplasmic membrane in microorganisms); PO) postsynaptic ending; PS) pre-synaptic ending; PV) pinocytosis vesicles; SER) smooth endoplasmic reticulum; SG) secretory granules; SV) synaptic vesicles.

Disruptions of the cell membranes are regarded as one of the major biological effects of ionizing radiations [8, 9]. In view of the general ex-perience that the earliest biochemical manifestation of irradiation is an in-crease in the activity of certain enzymes, Bacq and Alexander [8] proposed in 1955 that the primary events responsible for cell death are disturbances of the permeability of certain intracellular structures, notably the mito-chondria and microsomes. The "membrane effect," defined as a disturb-ance of the barrier between enzyme and substrate in the irradiated cell, was demonstrated in Russula nigricans tissues by Bacq and Herve [10] in one of the first experimental works in this field. Later Bacq and Alexander developed a hypothesis postulating a release of enzymes and explaining the effect of metabolism on development of the radiation-induced damage in terms of disturbances of the permeability and other properties of the intracellular membranes [11].   This hypothesis stimulated intensive research.

At the same time it was stressed [12] that molecular monolayers situated at the interface between zones of different composition may perform a function comparable to that of the major cell membranes in determining the spatial organization of metabolic reactions.   The compounds capable of creating structuralized surface layers and a molecular heterogeneity of the protoplasm include lipids, proteins and certain protein complexes — lipoproteins, nucleoproteins, etc.   The existence of various membranes in the living cell is responsible for some of the specific qualitative effects of radiation on the cell.   The irreversible chemical and structural lesions of cell membranes and fine interfaces are an important element in the disturbance of the time-space coordination of metabolic processes.

In 1957 Pasynskii [9] discussed the general theory of these processes, which are closely related to the behavior of living cells as open systems. He pointed out that in studying the effect of radiation on living organisms one should pay particular attention to the general properties of the chemical processes taking place in open systems.   Cells, as open systems, can maintain a stationary concentration of various components by means of permeability or diffusion.   The stationary concentrations of the components of open systems depend not only on the rate constants of chemical reactions but also on the permeability and diffusion constants.   The role of these parameters in the theory of open biological systems reflects the close relationship existing between the organization of metabolic processes in space and time.   Radiation-induced disturbance of the stationary state can be compensated within certain limits by the open system itself, although in the case of more severe damage the system cannot attain a new stationary state and therefore breaks down [13].

One of the crucial aspects of the biological effect of radiations is that irradiation of organisms causes considerable physiological changes even at comparatively low doses.   Radiation-induced disturbances of membranes and monomolecular interfaces undoubtedly play a major role in these processes.   The destruction of even a few molecules in a mono- or bimolecular interface in the cell can markedly alter the permeability of the layer and upset the stationary state of enzymatic reactions in the cell.   Through the chain of metabolic processes, these primary changes may trigger a whole sequence of biochemical disturbances [14, 15].   Some authors have stressed the importance of studying at the molecular level the close relationship between radiation damage to fine intracellular structures and oriented interfaces and changes in the diffusion parameters and stationary state of the cell.   The use of models is of great assistance, especially in the study of changes in the permeability of various monolayers taken as prototypes of fine biological interfaces.   The importance of studying radiation-induced lesions of the membranes surrounding that unique structure, the nucleus, has also been stressed [16].

The role of radiation-induced damage to membranes and the metabolic processes associated with these within the general framework of radiobiological effects is also stressed by the structural-metabolic hypothesis [17, 18].

This chapter discusses some of the models used to study radiation damage to fine biopolymer interfaces and the disruption of cell membranes and some of the biochemical functions associated with these membranes by radiation.

RADIATION DAMAGE TO SYNTHETIC BIOPOLYMER INTERFACES

The theoretical concept of the importance of radiation-induced damage to fine interfaces in the cell within the general framework of radiobiological effects has spurred on the experimental study of this problem at the molecular level. The use of models is of particular value for determination of changes in the permeability of cell membranes because it facilitates study of the effect of radiation on permeability under different controlled conditions.

Briefly, the experimental models consist of fine, membrane-like interfaces of natural biopolymers capable of fulfilling in particular one of the major functions of natural membranes, namely, that of serving as a barrier in reactions between enzyme and substrate.

The purpose of these studies was to determine the effect of irradiating fine interfaces of various types on the rate of the reaction between an enzyme and substrate separated by these interfaces.

Pasynskii and Volkova [15, 19] studied the effect of radiation on permeability of fine interfaces using peroxidase and synthetic lipoprotein complexes or RNA complexes with a nucleoprotein interface of about eight molecular layers. The experiments with nucleoprotein interfaces in peroxidase suspensions showed that a marked increase in the enzymatic oxidation of ascorbic acid is induced by only one or two ionizations in the membrane-like ribonucleoprotein (RNP) layer, containing over 1,000 RNA molecules. The lipoprotein complexes (caprylic acid-peroxidase) proved resistant to an irradiation of 50—70 kr under the prevailing experimental conditions; the authors attribute this to the marked tendency of lipoprotein components to spread on interfaces. Recent data indicate that nucleoproteins may be components of nuclear membranes, cell walls, etc. [20]. Radiation-induced damage to the structure of nucleoprotein interfaces is apparently crucial in the spread of disturbances in the biochemical processes of the cell.

Further studies were made of the effect of radiation on structuralized interfaces composed of molecules of lipoproteins, phospholipids and nucleoproteins, since these are known to be major components of the cell membranes. These experiments represented the first combined study of the surface properties of the above biopolymers on direct irradiation of their monolayers on a Langmuir balance, and the effect of direct irradiation on their permeability in specially devised models [21]. The compression isotherms of the monolayers, characterizing the surface properties of the biopolymers, were constructed from the readings on the Langmuir balance. The permeability effects were gauged from the change in the rate of reaction between an enzyme (peroxidase) and a substrate (methylene blue), separated by the monolayers.

Comparison of the experimental data obtained showed that the barrier function of the biopolymer monolayer attains a peak and remains constant when packing of the molecules on the surface reaches the limiting density. The monolayers are then in a phase state which is transitional between condensed and solid and are therefore subject to considerable Van der Waals forces. The minimum permeability of the monolayer is obtained at the compression exerted by a force of 5 dynes/cm in the case of deoxyribo-nucleoprotein (DNP) [21] and about 11 dynes/cm in lipovitellin [22]. Here we are dealing essentially with artificial membranes which facilitate the study of various properties of interfaces, and in particular the effect of radiation on their barrier function. It was found that only a surface com-posed of oriented molecules of DNP and lipovitellin (i. e., a monolayer of such molecules) can serve as a barrier. Interfaces composed of non-oriented molecules lack this property.

The surface oriented molecules of the biopolymer proved to be highly radiosensitive. The barrier function of the DNP monolayer decreased by 40% after irradiation at a dose of 2 kr and vanished completely at a dose of 5 kr (Figure 6.2). It appears that the radiation-induced structural changes in DNP monolayers consist in the aggregation and cross-linking of DNP molecules [23]. These assumptions are supported by the finding that changes in the pH of the subphase — a major factor affecting the interaction between DNP components [24] — does not disturb the barrier function of DNP monolayers [21].

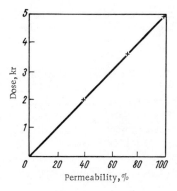

Figure 6.2. Effect of the dose of X-ray ir-radiation on permeability of DNP monolayers

In the case of DNP monolayers obtained from previously irradiated DNP solutions, an appreciable increase in permeability required doses as high as $5 \cdot 10^5$ to 10^6 r. Even then permeability of the monolayers increased by only 10—15%. Appropriate calculations and tests showed that these changes can be attributed to dissolution of part of the degraded SNP molecules in the subphase after irradiation of the solution. All other disruptions of the irradiated DNP molecules in solution appear to be corrected during formation of a DNP monolayer on a subphase of high ionic strength.

The exposure of a DNP solution to X-rays at a dose of 30—100 kr does not alter the mechanical properties of monolayers of the nucleoprotein subsequently formed [25].

UV irradiation of DNP monolayers at a dose of $2 \cdot 10^6$ ergs/cm² similarly accelerated the diffusion stage of the enzymatic reaction. In this case the barrier function of the monolayers decreased by more than 20%. An increase of the dose to $1.2 \cdot 10^7$ ergs/cm² reduced the barrier effect by 60% (Figure 6.3) [21]. Quantitative determination of the oxidation kinetics of methylene blue revealed a distinct change in the permeability of the UV-irradiated monolayers (Figure 6.4) [21]. The kinetics of the enzymatic reaction through a nonirradiated DNP monolayer does not follow Fick's law of free diffusion (see curve 1). UV irradiation at a dose of $1.2 \cdot 10^7$ ergs/cm² brought the enzymatic reaction rate closer to the rate characterizing free diffusion of the enzyme to the substrate (see curves 2 and 3).

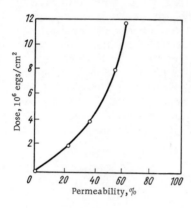

FIGURE 6.3. Effect of the UV irradiation on the permeability of DNP monolayers

Direct irradiation of DNP monolayers on the Langmuir balance reveals stable aberrations of the surface properties of this compound, manifested in an increase in the surface of the monolayer without a change in the course of the compression isotherms. The effect observed resembles the changes established [25] following interaction of a DNP monolayer with formaldehyde (dissolved in the subphase), which is known to induce intermolecular chemical bonds [26]. It is further known that such bonds can disturb the ordered structure of synthetic polymer films over a range of tens of atomic radii [27].

In natural interfaces, a considerable part of the lipid is bound with the protein into a solid "lyophobic core" [28, 29]. An example of such a natural lipoprotein complex is the lipovitellin contained in the yolk of hen eggs. Evans et al. [30], have shown that this lipoprotein contains a stable protein-lipid bond which is not affected by such factors as changes in ionic strength or pH, urea, or phospholipases C and D, and is only slightly influenced by detergents. Hence the interest in studying the effect of radiation on the surface properties of this natural lipoprotein and the barrier function of its monolayers [22].

Lipovitellin monolayers are more sensitive to direct UV irradiation than DNP monolayers. In this case a dose of $1.2 \cdot 10^7$ ergs/cm² completely eliminates the barrier effect of the monolayer and secures free diffusion of the enzyme to the substrate (Figure 6.5).

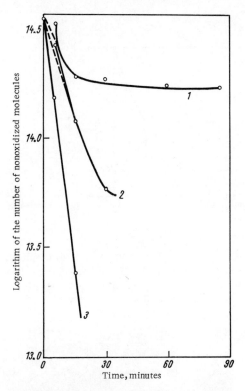

FIGURE 6.4. Kinetics of the enzymatic oxidation
of methylene blue in the coordinates of Fick's equation:

1) nonirradiated DNP monolayer barrier; 2) DNP monolayer barrier after UV irradiation at a dose of $1.2 \cdot 10^7$
ergs/cm^2; 3) free diffusion of the enzyme in the absence of a monolayer barrier.

The effect of radiation on the surface properties of lipovitellin monolayers is comparable to that obtained by a change in the ionic strength of
the subphase [31]. Moreover, the nature of change in the surface properties
of the UV-irradiated lipovitellin monolayer corresponds to the type of
changes observed in heat-denatured lipovitellin (the compression isotherms
almost coincide in these two cases). This may be due to weakening of the
hydrophobic interaction between protein and lipid, which causes reorientation of molecules on the surface. Electron micrographs of nondenatured
and heat-denatured lipovitellin confirm this hypothesis. Monolayers of
nondenatured lipovitellin show a much less distinct sequence of light and
dark bonds, corresponding to hydrophobic and hydrophilic parts of the
molecule, than denatured lipovitellin monolayers (Figure 6.6).

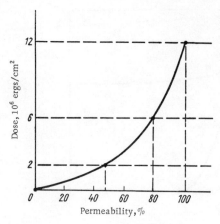

FIGURE 6.5. Effect of the dose of UV irradiation on the permeability of lipovitellin monolayers

The surface properties and barrier function of phospholipid interfaces are of great interest since phospholipids are major components of all natural membranes. Moreover, phospholipids stand out among the membrane lipids by virtue of their functional versatility [32, 33].

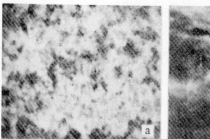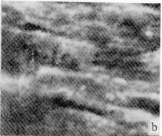

FIGURE 6.6. Electron micrograph of lipovitellin monolayers fixed with OsO_4 and stained with phosphotungstic acid:

a) nondenatured lipovitellin (300,000 ×); b) denatured lipovitellin.

The compression isotherms of phospholipid monolayers, which reflect their surface properties, differ markedly from those of DNP and lipovitellin. The surface molecules of phospholipids exist in only two phase states: gaseous and condensed. The absence of a solid phase is attributed [34] to the escape of part of the phospholipid molecules into the subphase at higher compressions of the monolayer. Irradiation of the monolayers on a Langmuir balance caused only a slight increase in their surface area [35].

Phospholipid monolayers possess no barrier properties under the conditions tested, but induce intensified enzymatic oxidation of the substrate. This effect becomes more pronounced on UV irradiation of the phospholipid monolayers at a dose of $1.2 \cdot 10^7$ ergs/cm^2 (Figure 6.7). Only surface-oriented phospholipid molecules, i. e., monolayers of the latter, show this property.

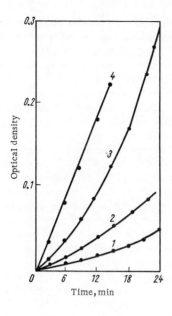

FIGURE 6.7. Kinetics of the enzymatic oxidation of methylene blue:

1) in the case of free diffusion of enzyme and substrate; 2) diffusion of enzyme through a layer of nonoriented phospholipid molecules; 3) diffusion of enzyme through a phospholipid monolayer; 4) diffusion of enzyme through an irradiated phospholipid monolayer.

Poltorak et al. [36, 37] previously reported an increase in activity of alkaline phosphatase and succinate dehydrogenase adsorbed on lipid layers oriented in this way.

It is known that lipids are fairly readily oxidized by molecular oxygen, yielding products of a varying degree of oxidation [38]. Peroxides appear to be the primary molecular products of the chain oxidation of lipids and hydrocarbons [39]. Enhancement of the enzymatic reaction of substrate oxidation following irradiation of phospholipid monolayers can also be explained partly in terms of peroxide formation in the hydrocarbon chains of the phospholipid molecules. The comparative study performed by Tongur et al. on the effect of UV irradiation on the behavior of saturated and unsaturated fatty acids in thin layers is of interest in this connection [40]. As in the case of phospholipid monolayers, irradiation of monolayers of palmitic and linoleic acids increases their area only slightly. Only monolayers of saturated fatty acid possess barrier properties. Monolayers of unsaturated fatty acid enhance enzymatic oxidation of the substrate under the experimental conditions used, though to a lesser extent than in the case of phospholipid monolayers. UV irradiation at a dose of $1.2 \cdot 10^7$ ergs/cm^2 increased this effect in the case of linoleic acid and induced it to a slight extent in the case of palmitic acid. It has also been shown that the activity of alkaline phosphatase adsorbed on the hydrophobic surface of a phospholipid (SiO$_2$ + kephalin) is almost twice as high as the activity of the same enzyme when adsorbed on the surface of a saturated fatty acid (SiO$_2$ + lauric acid) [36]. The radiation-induced disturbance of the barrier function of lipovitellin monolayers, reflected in the marked acceleration of enzymatic oxidation of the substrate [22], can apparently be attributed not only to a weakening of the bond between the lipoprotein components but also to the release of phospholipids from the complex.

Comparatively little information is available on the effect of ionizing radiation on the properties of monomolecular layers of proteins, which are nevertheless important components of all biological membranes. Attenuation of the serological properties of human serum albumin in mono- and bilayers under different conditions of irradiation with alpha particles and X-rays was observed by Rothen [41], Hatchinson [42] and Smith [43]. Mixed monolayers of pepsin and a substrate (egg albumin) stretched over an air/water interphase were found to be highly sensitive to X-ray irradiation at a dose of 25 r [44]. Later, however, these findings were seriously questioned [45]. In the case of UV irradiation of an egg albumin monolayer spread over an air/water interphase it was found [46] that low doses (from a low-pressure mercury lamp with 85% of its emission composed of light with a wavelength of 254 nm, range 15 cm, exposure 15 min) cause reorientation of the protein molecules on the surface owing to the rupture of hydrogen bonds. A further increase in the radiation dose sharply decreases the area occupied by the monolayer; this is due to degradation of the protein molecules and dissolution of their fragments in the subphase. Korgaonkar and Desai [47, 48] examined the properties of protein monomolecular layers in order to determine the effect of radiation on dissolved proteins. Aqueous solutions of histone, protamine sulfate, lysozyme and insulin (50 mg/ml) were exposed to gamma and beta radiation at doses ranging from $0.37 \cdot 10^5$ to $6.26 \cdot 10^5$ rad.

Irradiation markedly altered the properties of all the protein monolayers. In the case of histone and protamine an increase in the dose caused reduction in the area of the monolayer and altered the course of the compression isotherms, indicating disturbance of the mechanical properties of the monolayers. The molecular weight of these proteins in the monolayers changed from $15 \cdot 10^3$ to $3.5 \cdot 10^3$ in the case of histone and from $15 \cdot 10^3$ to $12 \cdot 10^3$ in the case of protamine — clearly an indication of molecular fragmentation caused by irradiation of the solutions. The overall decrease in the area of the monolayer is due to the escape of low molecular weight fragments into the subphase and the specific packing of the irradiated protein molecules at the surface.

Lysozyme and insulin showed the opposite effects at high radiation doses — namely, increase in the area of the monolayer and a rise in the molecular weight from $16 \cdot 10^3$ to $40 \cdot 10^3$ for lysozyme and from $12 \cdot 10^3$ to $22 \cdot 10^3$ for insulin. These changes indicate an aggregation of the proteins at high radiation doses. To gain a better understanding of the effect of radiation on protein molecules, a study was made of the surface properties of polyamino acids — poly-1-tyrosine and poly-dl-alanine — following gamma irradiation of their dilute solutions [49].

Since myosin and actomyosin-type proteins have been found in various cell membranes, the effect of irradiating aqueous solutions of myosin A on the properties of its monolayers deserves attention [45, 50]. It was found that exposure of dilute solutions of myosin A to low doses of X-rays (20 r) leads to a decrease in the area occupied by the macromolecule in the monolayer and causes partial neutralization of the charge. Moreover, irradiation lowers the elasticity and viscosity of the monolayer. This effect is enhanced by an increase in the protein concentration in the solution. The high radiosensitivity of myosin A, manifested in the surface denaturation

of the protein, is attributed to conformational changes in a small number
of molecules [45, 50]. These changes cause a marked disturbance of the
viscoelasticity and elasticity of the monomolecular protein layer in the
formation of its microstructure [45]. The exposure of dilute solutions of
myosin to comparatively high radiation doses causes fragmentation of the
macromolecules in solution; in concentrated solutions of myosin there is
an increased probability of intermolecular bonding, which leads to gel
formation and a concomitant increase in the elasticity and viscosity of the
monolayer. Large doses slow down formation of the microstructure of the
monolayer — an effect noted earlier by Tkach and Sidyakin [51]. These
authors examined the properties of monomolecular layers of serum proteins
(albumin and gamma globulin) following gamma irradiation of the protein
solutions using a Co^{60} source. They observed similar effects (over a wide
range of concentrations and at radiation doses from $30 \cdot 10^3$ to $270 \cdot 10^3 r$),
namely a reduction in the area of the monolayer as a whole and the area
occupied by each molecule, and increase in the compressibility and a slowing down of the rate of monolayer formation.

It follows from these data that irradiation of various macromolecules
either in oriented surface layers or in solution causes structural changes
which are reflected in the physicochemical, mechanical and barrier properties of the respective monomolecular layers. Such changes are possibly
the cause of radiation damage to structuralized and spontaneously arising
interfaces in the cell and may be responsible for both observed changes in
the diffusion parameters and for the shift in the trend of intracellular
metabolic reactions.

RADIATION DAMAGE TO CELLULAR AND
INTRACELLULAR MEMBRANES AND
MEMBRANE-ASSOCIATED BIOCHEMICAL PROCESSES

Research into the radiation-induced lesions of biological membranes is
being carried out not only in isolated structures but also at the cell and
tissue levels and in entire organisms. Only scant direct data are available
on radiation-induced changes in the permeability of membranes. This
scarcity of knowledge results largely from the limitations of experimental
study. Moreover, in analyzing data obtained from entire organisms it is
difficult to distinguish between the direct effect of ionizing radiations on
cell membranes and the inevitable indirect effects of total irradiation.
Nevertheless, the changes in permeability can be gauged indirectly on the
basis of physiological and biochemical data on the membrane systems of
the cell.

Permeability of cellular and intracellular membranes

The effect of ionizing radiations on permeability of membranes has been
studied in various animal and plant tissues. Kuzin and Strazhevskaya [52]

found that irradiation of wheat rootlets at a dose of 1,000 r markedly disturbs their permeability, causing considerable inhibition of the absorption of glycine C^{14} and phosphate $(Na_2HP^{32}O_4)$ within 30 and 60 minutes, respectively. During the first minutes (4 and 30 minutes) after irradiation (1 and 10 kr), excretion of radioactive phosphorus from irradiated roots is 3—4 times greater than from nonirradiated ones. These authors point out, however, that the observed changes vanish quite quickly (after 4 and 24 hours from the time of irradiation).

According to Budnitskaya et al. [53, 54], irradiation of 10-day old sprouts of various plants (cereals, legumes) at doses ranging from 1 to 50 kr alters the permeability of the tissue to electrolytes. The decrease in dispersion of electrical resistance occurring as early as 15 minutes after irradiation at a dose of 25 kr reflected a profound change in the structural organization of the leaf tissue. The loss of electrolysis and amino acids from gamma-irradiated carrot tissues in an aqueous medium depends on the dose and intensity of irradiation. Calcium ions escape most readily. These changes were detected immediately after irradiation at a dose of 10 kr [55].

Hluchovsky and Srb [56—59] observed early changes in permeability of plasma membranes, revealed by the plasmolysis of epidermal cells of irradiated and nonirradiated bulbs of Allium cepa L. Using both electrolytes (0.5 and 1 M KNO_3) and nonelectrolytes (sucrose, glycerol) as plasmolytic agents at pH 5, they detected two types of response: electrolyte-induced plasmolysis of the irradiated cells took place more slowly than in the controls, whereas in the presence of nonelectrolytes plasmolysis was more rapid than in the controls. These authors did not observe a dose effect, but it is noteworthy that changes in permeability were apparent at a dose as low as 25 r. On the other hand, Vasil'ev [60] did not detect any changes in tissue permeability of irradiated plants.

Brinkman et al. [61, 62] found that permeability of the skin, aortic wall and thin model membranes of connective tissue with respect to water and salt solutions increases immediately after irradiation at low doses. The mucopolysaccharide matrix of these objects showed immediate depolymerization.

It is evident from these findings that irradiation affects permeability of the cell membranes of various tissues, although it appears that not all experimental procedures can detect such changes.

We attach particular importance to the application of methods which reveal permeability changes during irradiation and facilitate a kinetic study of the process.

The effect of radiation on permeability of intracellular membranes is still little known. Scaife and Alexander [63] did not detect any change in permeability of mitochondrial membranes to nicotinamide adenine dinucleotide (NAD) following in vivo and in vitro irradiation at a dose of 9 krad. This can be attributed at least in part to the procedure used. The highly sensitive histoenzymatic-chemical reactions available today can detect enzyme activity at the tissue level and also early changes in permeability of mitochondrial membranes [64]. Moreover, the specificity of the change in membrane permeability to a particular chemical agent as a result of irradiation must be taken into account [65]. Some of the data available nevertheless show that radiation does affect permeability of mitochondrial membranes [66, 67].

Of particular interest is the permeability of nuclear membranes isolated from irradiated organisms or irradiated in vitro [68]. These experiments were performed on isolated rat thymus nuclei at different intervals (0.5, 1 and 2 hours) after irradiation at a dose of 1 kr. Permeability of the nuclear membranes was determined from the rate of penetration of glucose-C^{14} into the control and test nuclei, as well as from oxidation of reduced NAD by the nuclei and the escape of nucleotides and proteins from the irradiated and control nuclei. Irradiation increased the permeability of the nuclear membranes on the basis of these criteria, the effect being more pronounced in vitro than in vivo.

Lysosomes deserve particular attention in this respect. It is well known that the functioning of lysosomes under both normal and pathological conditions depends largely on their membranes. Irradiation in vivo [69−71] and in vitro [72−75] causes release and activation of lysosomal enzymes. Similar data were obtained with lysosomes of tumor tissues of irradiated animals [76].

Lysosomal structures were found in a number of cases to be highly vulnerable to irradiation in vitro. Since any indirect physiological effects can be ruled out in these cases, there is no doubt as to the damaging effect of radiation on lysosomal membranes. Wills and Wilkinson [74] observed a release of acid phosphatase, cathepsin and β-glucuronidase from rat liver lysosomes after irradiation in vitro at a dose of 1 krad. The effect was most pronounced in lysosomes kept at 4 or 20°C for 20 hours after irradiation. Similar changes were reported by Desai et al. [73] following irradiation at a dose of 2−10 kr [73]. However, Sottocasa and his colleagues [77] failed to detect any release of β-glucuronidase or β-galactosidase from a combined fraction of heart mitochondria and lysosomes irradiation in vitro at a dose of 440 kr. This may be due to a difference in affinity between the enzymes and the lysosomal membranes, or to some kind of structuralization of these particles. The actual biochemical processes triggered by a membranal lesion may obviously differ from one cell or lesion to another [11].

The adverse effect of radiation on membrane structures may be due largely to the peroxides generated by the primary radiochemical processes of lipid oxidation. Wills and Wilkinson [78], who examined the formation of lipid peroxides in irradiated suspensions of lysosomes, regard these compounds as an important factor in the release of lytic enzymes by the lysosomes [74]. It was found that the oxidation of fatty acids to peroxides causes destruction of membranes and leakage of hemoglobin from erythrocytes [79].

Apart from the direct action of radiation and water radicals, the membranes of irradiated tissues are also affected by radiotoxins created by disturbances in various metabolic processes. Thus, for example, lysosomes are damaged less when irradiated in vitro than when irradiated within tissue sections [77]. However, the increase in enzymatic activity in the tissues of irradiated animals is not always due to release of enzymes from the lysosomes. The increased activity of enzymes in the spleen and thymus is attributed by some workers to destruction of lymphoid tissue [80] or to an enhanced activity of hormones [81].

The escape of various compounds from irradiated cells

The escape of various soluble compounds, including proteins and nucleotides, from irradiated cells and organelles is discussed in the literature.
Although the mechanism of this phenomenon remains obscure, many workers
attribute it to membranal lesions. A number of publications, most of them
from Pollard's laboratory, contain data on the leakage of nucleic bases [82,
83], nucleosides and nucleotides [83–86], as well as β-galactosidase [87]
and other compounds from irradiated strains of E. coli. These findings
possibly reflect changes in the permeability and "rigidity" of the intracellular partitions. However, another interpretation of the observed effects
is also possible, as communicated in later works [83].

Various animal tissues have been used to study diffusion of the main
cofactors and enzymes through mitochondrial, nuclear, and plasma membranes. An escape of catalase [88], DNase [89, 90], RNase [91], NADase [92]
and proteolytic enzymes from the mitochondria was detected. Thus, the
exposure of an epithelial tumor in situ to X-rays at a dose of 1,000 rad
caused a 22% decrease in catalase activity in the mitochondrial fraction and
a corresponding increase in the cell fluid fraction [88]. Release of catalase
from the mitochondria of the epithelial tumor and migration of the enzyme
to the soluble cell fraction can be detected as early as 1–15 minutes after
irradiation at a dose of 50 rad [88, 94]. These changes, however, are highly
typical of tumor tissue membranes [94]. It was shown recently that the
mitochondrial membranes of tumor cells differ from those of normal cells
in various structural and functional aspects [95].

In a comparative study of the dynamics of DNase activity in the cytoplasm and nucleus, Argutinskaya and Salganik [96] found that activation of
the enzyme in the cytoplasm precedes its appearance in the nuclei. Thus,
DNase activity in the cytoplasm increases markedly as early as 2 hours
after exposure of the animals to X-rays at a dose of 1.2 kr, whereas incubation of nuclei isolated at the same time reveals practically no breakdown
of DNA. According to a number of workers, the increased activity of DNase
in the cell nuclei can be attributed to penetration of the enzyme from the
cytoplasm rather than to activation of nuclear DNase.

Data have been published on the loss of various enzymes by the nuclei –
namely, catalase [97], glycolytic enzymes [98] and NADase [99]. According
to Creasey [97], catalase activity in the nuclei of thymus cells decreases
to half the control value following total irradiation of the animals at a dose
of 1,000 r. Hagen and his colleagues [98] found a decrease in activity of
certain glycolytic enzymes in the nuclei after irradiation of the animals at
a dose of 700 r. There are two stages in the development of the radiation
effect: 1) the escape of enzymes from the nucleus into the cytoplasm, apparently via radiation-induced lesions of the nuclear membrane (during the
first two hours after irradiation); 2) inactivation of the enzymes within the
nucleus (6–8 hours or more after irradiation).

On the other hand, Japanese workers [100] have found that release of
lactate dehydrogenase from thymocytes irradiated at a dose of 1,000 r takes
place not immediately but only 3 hours after incubation of the cells in buffer
at 37°C (apparently coinciding with the death of the cells).

Irradiated nuclei also lose other proteins [68, 98, 101]. Thus, Ernst [101] observed depletion of the globulin fraction and histones in thymus, spleen and liver nuclei 30 minutes or more after irradiation of the animals at a dose of 1,000 r.

A decrease in the amount of NAD in thymocyte nuclei was reported by Scaife [92]. The loss of NAD was initially regarded as the first link in a chain of metabolic events [102]. However, Campagnari [103] found that the loss of NAD is a very late sign of cell damage and does not cause the death of thymocytes but rather accompanies it. The NAD content of thymus and spleen mitochondria decreases 4 hours after irradiation of the animals at a dose of 800 rad [92, 104]. Irradiation of animals at a dose of 5 kr or more causes similar changes in liver and tumor tissues [105, 106].

It is noteworthy that the content of reduced nicotinamide adenine dinucleotide phosphate (reduced NADP) in liver tissues [67] and their mitochondria [107, 108] decreases during the first few hours (0–3 hours) after irradiation of the animals at a dose of 800–1,000 r. The decrease in concentration of reduced NADP in the liver tissue of irradiated animals can be attributed largely to swelling of the mitochondria [109]. The reduced pyridine nucleotides are oxidized under such conditions and escape from the mitochondria into the cytoplasm, where they are broken down by dephosphorylation [110].

A deficiency of cytochrome c develops in the mitochondria of radiosensitive tissues after irradiation [104, 111–114]. The cytochrome effect is absent in heart and liver mitochondria [115, 116]. It is not known whether the changes observed in radiosensitive cells result from inactivation of cytochrome c, weakening of the bond with lipids, or increase in membrane permeability.

Of particular interest is the loss of cytochromes from nuclei isolated from irradiated animals or irradiated in vitro [117]. The amount of cytochromes b and c in thymocyte nuclei decreases markedly as early as 30 minutes after irradiation (1,000 r). This phenomenon reflects disturbance of the permeability of the nuclear membranes, apparently as a result of changes in the physicochemical properties of the membrane components.

Little is known about the effect of radiation on the physicochemical properties of membranes. Some tissues show a change in the sorptive properties of the cell surface [118]. The physicochemical changes in the mitochondria of irradiated cells, which may be assessed from the shift in their isoelectric point, impair the sorptive properties of these structures. Marked changes in the isoelectric point of the ribonucleoproteins of cell mitochondria can be detected within 5 minutes after irradiation of the animal at a dose of 1,000 r [119, 120].

Irradiation alters some of the major electrical properties of the cells, probably by damaging the cell membranes [121–124]. This explanation is supported by electron microscopic studies of the structure of irradiated erythrocytes [121, 125]. Polivoda [126] has established a relationship between the capacitance of the membranes of irradiated cells and the concentration of potassium in the medium. He suggests that radiation-induced damage to the cell membranes is associated with the strength of the electric field of the membrane [127].

Experiments with erythrocytes have shown that irradiation triggers a reaction between membranal components, possibly proteins, and iodoacetamide present at the time of irradiation. Such an interaction can be attribute to configurational changes in membranal proteins [128]. Tribukait has described an interesting phenomenon – rhythmic changes in the electrophoretic mobility of erythrocytes with an increase in the radiation dose (10 r or more) [129].

Changes in the physicochemical properties of biological membranes apparently play an important part in the effect of radiation on the cell as a whole. The above data show that these changes appear soon after irradiation and are often induced by quite low doses. These disturbances are possibly responsible for the more severe subsequent lesions of the membranal structure, ultimately leading to an increase in permeability, the escape of various cofactors and enzymes from the cell, and possibly inhibition of the capacity of the cell for repair [130].

Ionic permeability

The effect of radiation on cell membranes can also be assessed by examining the ionic permeability of the cell. Such studies of the surface membranes of cells have been made at the cell [131—142], tissue [143—146] and organism levels [147—148]. Data on disturbances of the ionic permeability at the subcellular level have also been published [149, 150].

Particular attention has been devoted to the effect of ionizing radiations on the permeability of erythrocyte membranes to cations. These works have shown that irradiation causes accumulation of sodium and escape of potassium from the cell [136, 151—153]. The observed disturbance of the ionic equilibrium of the cell possibly results both from damage to the active transport of ions associated with inhibition of the sodium pump [135, 154, 155] and from an increase in the passive permeability to cations, due in turn to structural changes in the membrane [128, 136, 137, 139, 156, 157].

According to Bresciani and his colleagues [154, 155], the exposure of human erythrocytes to X-rays inhibits both active transport of ions and specific ATPase involved in this process. It was found that active transport of sodium through the erythrocyte membrane is more sensitive to radiation than passive permeability. The authors found a difference in the radiosensitivity of the two components of erythrocyte membrane ATPase. The first component, which is inactivated by magnesium ions and is insensitive to sodium, potassium and cardiac glycosides, is inhibited by about 31% on irradiation at a dose of 8,900 rad. The second component, which is inhibited by ouabain and requires magnesium, sodium, and potassium for activity, is more radiosensitive since its activity is inhibited by 48% and 70% at a dose of 890 and 4,450 rad, respectively. The inhibition curve of the second component coincides with the inhibition curve of the active transport of sodium in intact erythrocytes. Analysis of dose-effect curves for the $K^+ - Na^+$ ATPase activity suggests that the membrane contains two parallel biochemical mechanisms transporting sodium with the participation of ATP, characterized by different radiosensitivities [155].

Differences in the radiosensitivity of various components of the erythrocyte membrane have also been reported by other workers [141]. Since
active transport of sodium is essential for maintenance of a physiologically
high concentration of potassium and low concentration of sodium inside the
cell against the opposite ratio prevailing in the medium [158], inactivation
of this transport by radiation may have lethal consequences for the cell.

However, Myers et al. [138, 159] adhere to a different viewpoint. They
propose that the observed overall change in the concentrations of sodium
and potassium in irradiated erythrocytes can be attributed to increase in
the passive flow of ions rather than decrease in the active transport. Experiments with isolated stroma [160] have shown, for example, that ATPase
is quite radioresistant. The most radiosensitive enzyme is acid phosphatase, which was appreciably inhibited even at a radiation dose as low as
4 krad. However, an increased loss of potassium from irradiated erythrocytes (10 and 60 kr) was detected during prolonged incubation at 4°C even
in the absence of any effect on the membrane ATPase [138, 159]. The observed loss of potassium was practically independent of the concentration
of cells during irradiation or the composition of the medium. Hemolysis
and lipid peroxides appeared at much higher doses (about 50 kr). On the
basis of these findings the authors conclude that escape of potassium from
the cells can be attributed to radiation damage to the membrane proper,
possibly on its inner side [138].

There are indications that the change in permeability of erythrocytes to
Na^+ and K^+ results from a radiation-induced change in the membrane
sulfhydryl groups [136, 137, 161]. The sulfhydryl groups responsible for
the radiation-induced change in permeability to Na^+ and K^+ appear to be
located both superficially and within the membrane. This is suggested by
the results of experiments with compounds which bind sulfhydryl groups
and either do or do not penetrate into the cell, as well as experiments involving enzymatic treatment of erythrocytes [128, 136, 137]. The alteration
of membrane sulfhydryl groups during irradiation apparently results from
a highly specific process in the membrane. These changes are specific to
monovalent cations; the rate of entry of anions, nonionic compounds and
bivalent cations is not affected by irradiation [137]. It appears that the
membrane SH groups possess a different radiosensitivity and that the
observed effect is due to a change affecting only a small number of sulfhydryl
groups [137, 156]. Sutherland and Pihl [156] examined in great detail the
effect of various factors on the radiation-induced modification of sulfhydryl
groups. Their experiments show that factors affecting the membrane structure can alter the number of radiosensitive SH groups. This may explain
the greater loss of SH groups on irradiation of isolated "ghosts" in comparison with intact cells. According to the above authors, the radiation-
induced loss of SH groups results from formation of disulfide, while restoration of the permeability [128, 136, 138] after irradiation is due to reduction
of these disulfides under conditions of an active metabolism.

The high radiosensitivity of passive transport is evident from a recent
study by Japanese workers [139], who showed by means of a new procedure
that the absorption of ^{22}Na by erythrocytes at 0°C is enhanced as a result
of gamma irradiation at a dose of 100 r or less. The increase in absorption

was partly inhibited by incubating the cells at 37°C after irradiation. Contrary to other sources [156], however, these authors did not associate the restoration process with metabolism. In their view, the radiation-induced changes in permeability concern the membrane proper, whereas the partial restoration which takes place at 37° can be attributed to rearrangements within the membranes.

Of particular interest are the data of Myers and Slade [128]. In these experiments, erythrocytes were irradiated in the presence of iodoacetamide and the radiation damage to the membranes assessed from the escape of K^+ from the cells, rate of hemolysis, formation of lipid peroxides, and sensitivity to snake venom. It is known that iodoacetamide normally reacts with SH groups; however, if the cells are exposed to radiation in the presence of this radiosensitizer, the latter can apparently react with any component of the membrane, as can be seen from the results of a number of model experiments [128, 157]. The above authors assume that irradiation causes a multitude of minor structural disturbances in the membrane, which are stabilized by iodoacetamide present at the time of irradiation. These changes are manifested during subsequent incubation more rapidly than in control experiments without the radiosensitizer. The authors believe that iodoacetamide reacts with membrane proteins and, though not affecting the strength of the membrane, disturbs its lipoprotein structure in such a way that the phospholipids become accessible to hydrolysis by snake venom phospholipase [128, 157].

In this connection experiments carried out in our laboratory deserve attention [162, 163]. It was found that irradiation of horse serum and serum albumin in the presence of labeled methionine enables detection of early denaturational changes which cannot be revealed by other known procedures. Thus, an increase in the binding of ^{35}S-methionine by serum albumin was observed after irradiation at a dose as low as 400 r. Here the methionine present at the time of irradiation similarly appears to reveal changes affecting the protein molecule usually manifested only in the presence of additional factors.

Somewhat contradictory data have been published with respect to the changes in ionic composition of cell nuclei on irradiation. Creasey [150] found that the exposure of rat spleen and thymus to X-rays and gamma rays both in vivo and in vitro causes an escape of sodium and potassium from the nuclei. In the in vitro experiments, loss of potassium and sodium from spleen nuclei was detected as early as 5 minutes after irradiation, starting from a dose of 5 and 42 r, respectively. Several years later, however, Jackson and Christensen [164] repeated the same experiments and were unable to confirm these findings.

The increased concentration of magnesium in thymus nuclei, observed a short time (3 minutes or more) after exposure of rats to radiation at a dose of 1,000 r, has been tentatively attributed to disturbance of the permeability of the nuclear membrane [165].

The mitochondria — the organelles storing most of the energy in the cell — play a major role in the active transport process. They can also accumulate Ca^{2+} [166–168], Mg^{2+} [169], K^+ [169, 170] and Mn^{2+} [171] against a concentration gradient, and the migration of these cations inside the mitochondria follows certain general principles [169].

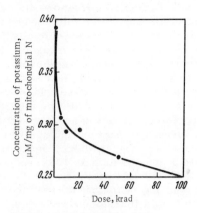

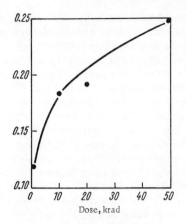

FIGURE 6.8. Concentration of potassium in rat liver mitochondria after an in vitro irradiation. The mitochondria, irradiated in 0.25 M sucrose, were incubated at pH 7.4 in 5 mM Tris buffer with 10 mM KCl for 30 minutes at 25°C [149]

FIGURE 6.9. Loss of potassium into the incubation medium by irradiated rat liver mitochondria. The mitochondria were irradiated in sucrose and were then incubated for 30 minutes in a potassium-free medium at 25°C. (The ordinate axis shows the loss of potassium in μM per mg of mitochondrial N.) [149]

Only a few works deal with the effect of ionizing radiations on ion transport in mitochondria. Wills [149] found that the processes involved in the transport of K^+ and Ca^{2+} in rat liver mitochondria are sensitive to relatively low radiation doses (5 krad) in vitro. Irradiation of isolated mitochondria with electrons at a dose of 2–100 krad and subsequent incubation in the presence of a low concentration of potassium (10 mM) or without the latter leads to a loss of K^+ into the medium (Figures 6.8 and 6.9). Absorption of ions by the irradiated mitochondria is enhanced by the presence of a higher concentration of potassium in the medium (0.1 M). In addition, irradiation slows down the active consumption of calcium ions and accelerates the loss of endogenous Ca^{2+}. Wills [149] assumes that the observed changes can be attributed to lesions of the mitochondrial membrane and are not associated with inhibition of ATPase or oxidative phosphorylation (OP). Such is also the conclusion of Gorizontova and co-workers [65], who observed increase of the accumulation of ^{54}Mn ions by liver mitochondria 24 hours after irradiation of the animals with a 800 r dose.

The adverse effect of radiation on the ion transport has also been demonstrated in plant mitochondria. However, these changes were detected long after irradiation. Thus, Merkulov [172] found a marked decrease in the efficiency of absorption of calcium ions by the mitochondria of 4-day old maize shoots grown from seeds exposed to a 50 kr dose of gamma radiation.

Kalacheva [173] examined the substrate-dependent absorption of calcium ions by mitochondria isolated from normal 12-day old green pea seedlings exposed to X-rays at doses of 1, 5, and 10 kr. Figure 6.10 shows the effect of the radiation dose on absorption of calcium ions by the mitochondria. Irradiation at a dose of 1 kr causes immediate activation of ionic absorption, followed by sharp inhibition of the process as the dose is increased. At the same time the escape of calcium ions from the mitochondria is enhanced at all doses tested (1, 5, and 10 kr). As shown in Figure 6.11, loss of calcium ions from the mitochondria increases with the radiation dose. Increase in the loss of Ca^{2+} at doses of 1 and 5 kr is respectively 0.0381 and 0.0960 μ M per mg of mitochondrial N. These data reflect the increased permeability of the mitochondrial membrane. They agree with earlier findings showing that when the P/O levels in the control are high, the efficiency of OP can only be reduced by exposing the plants to a dose as high as 10 kr, and even then the reduction is slight [174]. At the same time the radiation effect is enhanced by washing the mitochondria or preincubating them in the absence of the oxidation substrate. This suggests that the radiation-induced reduction in OP may be due to initial disturbance of the structure of the mitochondrial membrane.

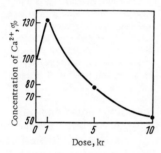

FIGURE 6.10. Effect of radiation on the absorption of calcium ions by the mitochondria of pea seedlings immediately after irradiation in vivo (relative to the control)

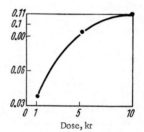

FIGURE 6.11. Increase in the loss of calcium ions by the mitochondria of pea seedlings as a function of the radiation dose (relative to the control). The values on the ordinate axis represent the loss of Ca^{2+} on irradiation in μM per mg of mitochondrial N

It may be assumed on the basis of experimental evidence that disturbance of Ca^{2+} transport in the mitochondria of irradiated plants, comprising decrease in the ionic absorption and enhancement of the escape of the ions, results, in the same way as the change in OP, from damage to the physicochemical structure of the membrane components responsible for the "passive" permeability of the membrane.

It appears, therefore, that loss of the ion-retaining capacity of the mitochondrial structure exerts a crucial influence on the course of the fundamental biochemical processes within this organelle, and that the resulting increase in the intracellular concentration of ions can bring about a series of metabolic disturbances in other subcellular components. However, the molecular mechanisms of ion transport have not yet been fully clarified, and the existing concepts will have to be corrected as fresh evidence becomes available.

The energy metabolism of the cell

ATP is formed exclusively on lipoprotein membrane structures of the cell. The most highly perfected structures as regards the energy-generating process of OP are the autochondria. These structures have been studied in greatest detail both by biochemical procedures and by electron microscopy. It was established not long ago that the respiratory chain and the OP system are associated with the inner membrane [175—178]. The relationship between the structure and metabolic function of the mitochondria has been widely discussed in the literature [179—181]. Experimental evidence concerning the effect of ionizing radiations on mitochondrial OP will therefore be discussed with reference to the close relationship between metabolism and structure.

Numerous publications show that irradiation of animals causes a decrease in utilization of inorganic phosphate for synthesis of energy-rich bonds during incubation of mitochondria isolated from their tissues — in other words, there is an uncoupling of OP [182—185]. These functional changes, accompanied by structural damage, appear in some cases within 30 minutes to 4 hours after exposure to a lethal dose of radiation. Disturbances of OP have been established even in the mitochondria of the relatively radioresistant liver tissues at various intervals after irradiation and different oxidation substrates [186—190], as well as in yeast cells [191] and plant mitochondria [174, 192].

These data clearly demonstrate that radiation does exert an adverse effect on OP, though as a rule not immediately and only if lethal doses are applied. Some workers, however, deny that radiation-induced uncoupling of OP takes place in radioresistant and even in radiosensitive tissues [193, 194]. This may be attributed at least in part to the use of different procedures. We attach particular importance to the choice of test for detection of damage. The capacity of mitochondria for coupled OP is commonly measured by such criteria as the respiratory control (RC), P/O, and the ATPase activity. RC is a more sensitive index than the coefficient P/O. Nevertheless, the most commonly used criterion is P/O.

Since any biological system involves constant competition between delivery of energy-rich compounds and the various pathways of their utilization, it is necessary to determine not only the coefficient P/O corresponding to the thermodynamic efficiency of oxidation, but also the kinetic efficiency, which is measured by the phosphorylation rate and is equivalent to the output. Recent studies in this field have used rapid recording techniques for measuring kinetic efficiency [195, 196], in contrast to the manometric period of research, during which the efficiency of substrate utilization was assessed mainly by the thermodynamic coefficient P/O.

According to Romantsev and his colleagues [197], the functional state of the liver mitochondria of rats exposed to gamma radiation at a dose of 1,000 r isolated from the liver 3 minutes, 30 minutes and 3 hours after irradiation, corresponds to "loose" coupling of the oxidation-reduction processes in contrast to the "firm" coupling characteristic of the liver mitochondria in healthy animals. Determination of R/C is impossible under such circumstances. After 30 minutes there is a change in the ultrastructure of the organelles correlated with modification of the OP processes.

Similar changes in liver mitochondria were observed by Yang Fu-yu et al. [184, 185]. A decrease in the RC, accompanied by swelling of the mitochondria, became evident as early as 30 minutes after the exposure of mice to X-rays at a dose of 700 r and increased over the course of one hour, whereas decrease in the coefficient P/O was negligible and was detected only one hour after irradiation. The changes in structure and function of the mitochondria were temporary and disappeared completely within four hours after irradiation. Restoration of the initial RC in liver mitochondria irradiated in vivo was also observed by other authors [197]. Similar data were obtained with plant mitochondria [198].

Some publications stress the greater radioresistance of plant mitochondria [199], especially those of fruit tissues, which have a comparatively sluggish metabolism [198, 200, 201]. In a series of investigations, the Californian school [198, 200, 201] has demonstrated the great structural and functional stability of mitochondria isolated from the tissues of irradiated fruits. Thus, a reversible change in the RC soon after irradiation appears only at doses as high as 250—500 krad. This can be attributed to a high capacity for repair on the part of the organelles, since the cells of metabolically sluggish tissues evidently have limited energy requirements, and coupling between oxidation and phosphorylation processes increases with ripening of the fruit [202, 203].

Despite prolonged research into the processes of OP, the mechanism of the uncoupling action of ionizing radiations remains largely obscure. Evidently, the radiation-induced disturbance of coupled OP cannot be due to enhancement of mitochondrial ATPase activity [183, 187, 204].

Disturbances of mitochondrial structure and function occur in the tissues and organs of animals exposed to total irradiation; isolated mitochondria, on the other hand, are markedly radioresistant [104, 111, 182]. This suggests that modification of OP in the irradiated organism is enhanced by disturbances of hormonal regulation [205] and the formation of lipid peroxides [206, 207] or other quinoid-type radiotoxins [18, 208].

Inhibition of OP in irradiated cells is associated with structural lesions of the mitochondria. It is known that the number of mitochondria per cell varies from only a few to 500,000 in different organisms [209]. As a rule, the cells of radioresistant tissues contain numerous mitochondria. For example, a rat liver cell has about 1,000 mitochondria, whereas radiosensitive cells such as spermatogonia or lymphocytes are relatively poor in mitochondria. The existence of a distinct correlation between the number of mitochondria per cell, a low level of OP and the radiosensitivity of cells has been confirmed by Goldfeder and his colleagues [88, 210]. The number of mitochondria and their structural and functional activity are regarded as important factors not only with respect to the initial radiation effects but also in the repair processes [88, 211, 212].

Early morphological changes in the mitochondria of radioresistant tissues exposed to in vivo irradiation comprise swelling, clarification of the matrix and destruction of the cristae [66, 109, 184, 197, 213—215].   In a study of early changes in the cytoplasm of liver cells of mice and rats 30 minutes after exposure to X-rays (at doses of 150, 300 and 450 r in the case of mouse cells and 800, 3,200, 6,400 and 9,600 r in the case of rat cells), Braun [214] found that the mitochondria show a tendency to condense and the ergastoplasmic lamellae are reduced.

Disturbances of the fine structure of the mitochondria of radioresistant tissues are often reversible and disappear with time [184, 197, 216].   Irreversible structural lesions develop in radiosensitive cells soon after exposure to relatively small radiation doses [211, 213].   According to Manteifel' and Meisel' [211], the mitochondria of the lymphatic node lymphocytes of rats exposed to X-rays at a dose of 500 r show distinct changes, including swelling and often fusion of the organelles, destruction of the cristae and clarification of the matrix, as early as 30 minutes after irradiation.   Ultrastructural changes appear in the lymphocytes earlier, 4—5 minutes after irradiation, at higher doses [212].   These disturbances affect the lipoprotein membranes of the cell organelles (mitochondrial cristae and membranes, nuclear membranes).   They comprise determination and vacuolization of the double-layered membranes and destruction of the mitochondrial cristae.   The mitochondria are affected first, then the nuclei. Similar results have been reported by Goldfeder [88].   Thus, 11 hours after exposure of mice to X-rays at a dose of 400 r, swelling of the mitochondria, destruction of membranes and fusion of ribonucleoprotein particles were observed in the lymphocytes.   Some of the nuclei and their membranes remained intact.   The mitochondria of irradiated animals are typically in close contact with the nuclear membrane [209, 211, 212].

There can be no doubt as to the biological importance of the nucleus for the cell.   The nucleus contains the genetic information of the cell and synthesizes nucleic acids, proteins and many other compounds essential to both cell and nucleus.   The main source of energy for these processes is ATP, which can be synthesized by glycolysis or by OP [217, 218].   The occurrence of OP in the nuclei was doubted in the past.   However, today it is well known that isolated nuclei of rat and calf thymus can perform an independent synthesis of ATP from intranuclear precursors under aerobic conditions [219, 221].   The first studies in this field were made by Allfrey and Mirsky [222, 223], who showed that nuclei isolated from calf thymus can phosphorylate the AMP present in them into ADP and ATP.

Nuclear phosphorylation is highly radiosensitive [224—228].   Creasey and Stocken [225] found that phosphorylation in the nuclei of the thymus and other radiosensitive tissues of rats is totally inhibited by irradiation at a dose of 100 r both in vivo and in vitro.   Later studies have confirmed the high radiosensitivity of nuclear phosphorylation.   However, some contradictory findings have also been published.   Repeating the work of Creasey and Stocken [225], Zimmerman and Cromroy [229] obtained a negative result: the nuclei did not lose their ability to form acid labile phosphate after irradiation. These authors stress another finding, namely, the escape of orthophosphate from the irradiated nuclei.

The high radiosensitivity of the intranuclear synthesis of ATP was clearly demonstrated by Klouwen and Betel [227]. Using various analytical methods, these authors detected inhibition of the nuclear synthesis of ATP in vivo in rats exposed to X-rays at a dose of 50—700 r. This effect was absent in vitro experiments involving irradiation of suspensions of nuclei, contrary to the findings of Creasey and Stocken. In a later publication Klouwen [230] reported a 26% decrease in the rate of formation of labile phosphate in the nuclei of thymocytes irradiated at a dose of 200 r; this effect, however, was not immediate but could only be detected after incubation for one hour at 37°C.

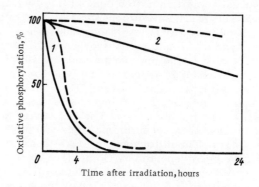

FIGURE 6.12. Effect of total irradiation (900 r) on OP (relative to the control). Average results obtained with thymus, myeloid leukemia and C57BL lymphosarcoma (1) and F_1 and CBA lymphosarcoma (2):

——— nuclear oxidative phosphorylation;
- - - - mitochondrial oxidative phosphorylation [231].

Radiosensitivity of nuclear and mitochondrial OP varies greatly from one cell type to another [231]. The authors of this study used three lymphosarcomas grafted on to mice — C57BL, CBA, and F_1 — as well as RF mice myeloid leukemia and thymus cells, the latter serving as controls (Figure 6.12). In one group of cells, inhibition of the synthesis of nuclear ATP was observed during the first two hours after exposure to radiation at a dose of 900 r. Mitochondrial phosphorylation in these cells was likewise inhibited at an early stage, but this inhibition apparently reached a peak some time after inhibition of nuclear ATP synthesis. In the other group of cells nuclear oxidative phosphorylation was inhibited later, between 8 and 24 hours after irradiation, apparently coinciding with the manifestation of cytological changes. Mitochondrial OP in these cells remained undisturbed right up to the time of cell disintegration.

On the basis of these results, the authors suggest that the pattern of biochemical changes is specifically linked with the type of cell death following

irradiation. The existence of such a relationship would explain the biochemical mechanism of interphase death. Thus, if ionizing radiation causes
rapid inhibition of both nuclear and mitochondrial OP, the cells die during
interphase as a result of energy depletion; on the other hand, if mitochondrial and nuclear phosphorylation is relatively slightly affected, the
cells survive until the next mitosis [231].

In a study of the inhibiting effect of radiation on nuclear synthesis of
ATP, Betel [232] found that the rate of ATP synthesis decreases not immediately following irradiation but only after incubation for 20 minutes;
the utilization of ATP is similarly unchanged. However, the concentration
of adenine nucleotides in both the cells and the nuclei decreases immediately
after irradiation. Betel regards the latter phenomenon as a possible cause
of the observed inhibition of ATP synthesis in thymus nuclei.

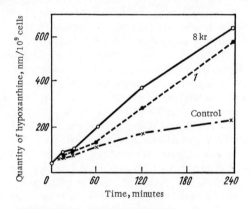

FIGURE 6.13. Liberation of hypoxanthine from
irradiated thymocytes into the incubation medium
as a function of the time of incubation [233]

Araki and his colleagues [233] tried to identify the degradation products
of adenine nucleotides in irradiated thymocytes. It is known that degradation of adenine nucleotides ultimately yields hypoxanthine. Indeed, liberation
of hypoxanthine into the medium was detected at an early stage in the two-
hour incubation of a suspension of irradiated thymocytes with glucose at
37°C (Figure 6.13). The amount of hypoxanthine liberated depended on the
dose of irradiation within the 0.5–32 kr dose range (Figure 6.14). These
findings confirm Betel's assumption that the rapid depletion of adenine
nucleotides in thymocytes, immediately after their irradiation, slows down
the synthesis of ATP. However, the mechanism of this process is still
obscure.

Kuzin and Tarshis [234] made a comparative study of the radiosensitivity
of ATP synthesis in thymus nuclei shortly (30 minutes) after irradiation
in vivo and in vitro at a dose of 1,000 r. Inhibition of ATP synthesis was
found to be more pronounced in vitro — a finding attributed to the action of
water radicals on the unprotected surface membrane of the nucleus.

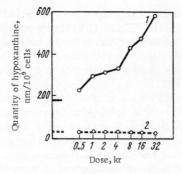

FIGURE 6.14. Accumulation of hypo-xanthine in the incubation medium of rat thymocytes as a function of the radiation dose after 2 hours of incuba-tion (1) and prior to incubation (2) [233]

All these data indicate that nuclear OP is a highly radiosensitive process in radiosensitive tissues, but is inhibited only 30—60 minutes after irradia-tion. Discrepancies in the results with respect to the radiosensitivity of this process in vitro are probably due to differences in the procedures used for isolation of the nuclei. However, the data available are as yet insuffici-ent to draw any definite conclusions as to the factors responsible for dis-turbance of the intranuclear synthesis of ATP. One possible factor is a change in the conformation of the membrane proteins, leading to alteration of the permeability and a marked loss of cytochromes [117]. Clearly, the loss of cytochromes [117] and decrease in the respiration rate [235] in ir-radiated nuclei reflect radiation-induced damage to the nuclear membrane.

The adverse effects of radiation on the nuclear membrane can also be assessed by determining the radiosensitivity of the various membrane components. We found, for example, that the ATPase of the nuclear fraction composed of nuclear membranes and acid proteins is much more radio-sensitive than the ATPase of intact nuclei. An appreciable inhibition of this ATPase was obtained immediately after irradiation at a dose as low as 10 kr (Figure 6.15). A comparable effect in intact nuclei required a dose of at least 100 kr (Figure 6.16) [236].

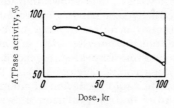

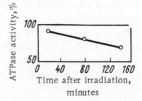

FIGURE 6.15. Inhibition of the ATPase activity of a nuclear fraction composed of nuclear membranes and acid proteins. The enzymatic activity was measured 20 minutes after irradiation [236]

FIGURE 6.16. Inhibition of the ATPase activity of an irradiated suspension of rat thymus nuclei as a function of the time after irradiation at a dose of 100 kr [236]

Biochemical studies have shown that isolated nuclear membranes of liver cells are rich in ATPase activity [237]. The presence of a contractile protein possessing ATPase activity inside or near the membrane of thymus cells was reported by Ohnishi [238]. Histochemical data obtained by electron microscopy indicate that ATPase is present in the pores of the nuclear membrane [239—242]. In view of the available data on membranal localization of nuclear ATPase, the partial inactivation of ATPase obtained at a dose as low as 10 kr in our experiments can be interpreted as evidence of direct damage to the nuclear membrane.

Electron microscopy has revealed radiation-induced lesions of nuclear membranes [66, 212, 215, 243, 244]. The morphological changes observed in the nuclei of thymocytes and lymphoid tissues include delamination of the two-layered nuclear membrane with the subsequent appearance and expansion of a perinuclear space; breaks in the periphery of the nuclear membrane also appear in some cases. These changes are discernible immediately after irradiation; it is noteworthy that the inner structure of the nuclei of these lymphocytes (4—5 minutes after irradiation of rats at a dose of 8,000 r) remains unchanged. The earliest morphological changes in interphase lymphocyte nuclei are found in the nuclear membrane [212], although in the cell as a whole the mitochondria show earlier and more pronounced changes.

Structural examination of the neurons of the superior cervical sympathetic node of rabbits exposed totally to X-rays at a dose of 1,400 r reveals a rapid reaction of the mitochondria and endoplasmic reticulum to irradiation. Among the changes observed are reduction of the small vesicles of the Golgi apparatus, convolution of the nuclear membrane and widening of its pores as well as of the perinuclear space. These changes have been described in an outline of the development of acute radiation disease [215].

The high radiosensitivity of the nuclear membrane was demonstrated by Nadareishvili and his colleagues [245]. In a study of the kinetics of the hypotonic and alkali-induced destruction of the nuclei of nerve and glial cells, these authors observed nuclear lysis following irradiation at a dose of 500—1,000 r. They suggest that irradiation causes latent lesions in the nuclear membranes which become evident as a result of additional factors. Nuclei isolated from various organs and tissues do not differ in radiosensitivity according to the criterion of osmotic stability — their dose-effect curves coincide [246].

The role of radiotoxins and histones in biomembrane lesions

The accumulation of lipid peroxides on irradiation can be regarded as one of the possible causes of the adverse effect of radiation on cellular and intracellular membranes. As long ago as 1954 Tarusov suggested that the evolution of radiation-induced damage depends on the appearance of fatty acid peroxides in the affected organs and tissues [247]. According to Tarusov and his school, the accumulation and formation of these compounds is a primary process which takes place as a chain reaction [248, 249]. In vitro experiments have shown that irradiation of fatty acids at high doses can yield oxidized derivatives of these compounds, the process being dose-dependent [248, 249].

Attempts have been made to induce radiation disease by introducing irradiated fatty acids into animals. According to Kudryashov [249], the introduction of lipid toxins into the organism causes pronounced deformation, swelling and aggregation of the mitochondria and sometimes even their total destruction. Lipid toxins, by damaging the structure of the mitochondria, cause functional changes in the energy system of the cells and accelerate the breakdown of proteins in the liver. This effect is directly proportional to the degree of oxidation of the unsaturated fatty acid.

It was also found that the tissues of irradiated animals contain lipid-type compounds capable of uncoupling the process of OP [250]. In rabbits exposed to severe irradiation, the acetone fraction of liver lipids shows an enhanced capacity to slow down OP and reduce the coefficient P/O when added to isolated mitochondria. Serum albumin activates OP in mitochondria treated in the above manner. The coupling effect of albumin shows clearly that the adverse effect of lipids from irradiated animals on the phosphorylation process is due to the presence of free unsaturated fatty acids among them.

Budnitskaya and Borisova [251, 252] found a marked increase in the concentration of lipid peroxides in the leaves of seedlings soon after irradiation at comparatively low doses. They attributed this effect to the increased activity of lipoxidase [253, 254]. The formation of lipid peroxides on irradiation and their adverse effect on the various membrane systems of the cell were reported by Wills and Wilkinson [74, 78, 255].

In an analysis of the mechanism of permeability Robinson [256] discusses the possibility of interaction between the sulfhydryl groups of proteins and unsaturated lipid bonds. The role of such interactions between specific proteins and lipids may be of particular importance in the membrane pores in connection with the regulation of permeability. The existence of such an interaction has been suggested as an explanation of the radiation-induced changes in the membrane [257].

Quinoid type radiotoxins play an important part in radiation-induced damage to biomembranes. More details on this topic can be found in the monograph by Kuzin [18].

No discussion of the radiation-induced lesions of intracellular membranes would be complete without consideration of the so-called free histones, generated by dissociation of the nucleoprotein complex under the influence of radiation. Data on this topic have been published. Thus, labilization and weakening of the DNA-histone interaction following irradiation was observed in several laboratories [258, 263]. There is also evidence of liberation of histone from the DNA of chromatin granules [264–266] and decrease in the histone content of the nuclei of plant [267] and animal cells [101, 268–272].

There is no doubt today that the function of histones is not only structural but also regulatory, because of their remarkable propensity for reacting with various biopolymers and other compounds in the cell. It is known that histones induce marked changes in the properties and structure of various biological membranes: they enhance the permeability of cell membranes

[273, 274] and nuclear membranes [275], they cause marked swelling of mitochondria [276, 277] and influence the mitochondrial transport of ions [278–280], accelerating in particular the escape of potassium ions into the medium.   Mitochondrial respiration and OP [281, 282], as well as ATP synthesis and respiration in thymus nuclei [219, 275] are stimulated by low concentrations and inhibited by high concentrations of histones.   Moreover, histones exert a major influence on nuclear glycolysis [219, 283–285].   Depending on their concentration and other conditions of the experiment, histones either stimulate or inhibit the activity of mitochondrial ATPase [286], $Na^+ - K^+$ ATPase from the renal cortex of guinea pigs [287], and actomyosin ATPase [288]; they inhibit the ouabain-sensitive heart ATPase [289] and various microsomal ATPases activated by magnesium ions [288], as well as the activity of cytochrome oxidase [290].   The activity of succinate oxidase is respectively stimulated and inhibited by histones at a concentration of 50 and 250 mcg [282].   Nuclear NADH oxidase is more resistant to inhibition by histones than mitochondrial NADH oxidase [291]. The effect of histones on the permeability of intracellular membranes and the catalytic activity of enzymes is discussed by Braun [292].

It is evident from this brief outline that histones exert a major influence on the system of intracellular membranes and the entire metabolism of the cell.   However, the cytological and biochemical changes resulting from liberation of histones only appear at least a full hour after irradiation. This means that the damaging action of histones can only be a secondary process aggravating the primary effect of radiation.

CONCLUSION

Studies of radiation damage to membranes at the cellular and subcellular levels, as well as in tissues and entire organisms and in model systems have demonstrated the great vulnerability of these structures to radiation.

Especially susceptible to radiation damage are the mitochondrial membranes, in which ultrastructural lesions become evident earlier than in other cell membranes, including the nuclear membrane.   Radiation damage to the nuclear membrane is nevertheless crucial in view of its general regulatory function in the cell and its connection with mitosis.

Recent works contain abundant experimental evidence concerning the effect of radiation on the biochemical properties and functions of various membrane systems.   The degree of membrane injury is commonly assessed by the change in permeability.   However, not all procedures can reveal this effect.   Experiments made in vivo and in vitro have shown the great vulnerability of lysosomal structures, expressed by the release and activation of lysosomal enzymes.   Of particular importance in the development of radiation-induced membrane lesions are lipid peroxides, the formation of which has been demonstrated in suspensions of irradiated lysosomes and other cell organelles.

Some authors believe that membrane lesions are responsible for the escape of various components from irradiated cells.

The damage to cellular and intracellular membranes is reflected in disturbances of the ionic permeability. Of particular interest in this connection is the enhancement of the escape of endogenous K^+ and Ca^{2+} from liver mitochondria exposed to relatively low doses of radiation (5 krad) in vitro.

Irradiation causes accumulation of sodium in the cells and escape of potassium from them. The passive permeability of erythrocytes to cations is highly radiosensitive.

The possible mechanism of the radiation-induced disturbance of permeability has been discussed. It is suggested that destruction of membrane SH groups performs an important role in this process. This is an attractive hypothesis since many membrane proteins, including enzymes, participating in the transport of substances across the membrane, contain sulfhydryl groups.

Physicochemical changes in membranes appear to be of particular importance in radiobiological phenomena. The disruption of weak intra- and intermolecular bonds, causing conformational changes in the membrane proteins, is especially crucial. However, experimental data in this field are practically nonexistent. Research into radiosensitivity of the allosteric centers of membrane enzymes may therefore be of particular interest. Detailed data are already available for individual enzymes in solution.

Valuable data on the radiosensitivity of membranes have been obtained by the application of additional factors including osmotic shock.

Of particular interest is the possible destruction of the lipoprotein structure of membranes at the moment of irradiation. Such changes in the membrane as a result of low doses of radiation (less than 100 r) cause loss of potassium by the cells. Our understanding of the vulnerability of membrane components has been enriched by model experiments. It was found, for example, that lipoprotein and phospholipid interface monolayers are highly radiosensitive. The radiosensitivity of other biomembrane components must be determined.

Research into the direct destructive effect of radiation on membranes at the molecular level and the relationship between these lesions and the earliest biochemical disturbances observed in irradiated cells will contribute much toward an understanding of radiation damage to cell membranes.

BIBLIOGRAPHY

1. Malhotra, S.K.— Prog. Biophys. Molec. Biol. 20 (1970), 69.
2. Wallach, D.F.H.— J. Gen. Physiol. 54 (1969), 3.
3. Lenard, J.— Trans. N.Y. Acad. Sci. 31 (1969), 872.
4. Vanderkooi, J. and D.E.Green.— Proc. Natn. Acad. Sci. U.S.A. 66 (1970), 615.
5. Sitte, P.— Ber. dt. bot. Ges. 82 (1969), 329.
6. Neifakh, S.A.— In: Mekhanizmy integratsii kletochnogo obmena. Edited by S.A.Neifakh, p.9. Leningrad, "Nauka," 1967.
7. Gel'man, N.S. et al. Bacterial Membranes and the Respiratory Chain.— Moscow, "Nauka." 1972. (Russian)
8. Bacq, Z.M. and P.Alexander. Fundamentals of Radiobiology.— London, Pergamon Press, 1955.
9. Pasynskii, A.G.— Biofizika 2 (1957), 566.
10. Bacq, Z.M. and A.Herve.— Bull. Acad. r. Med. Belg., 6th ser. 18 (1952), 13.

11. Bacq, Z.M. and P.Alexander. Fundamentals of Radiobiology.— London, Pergamon Press, 1955.
12. Pasynskii, A.G. Biophysical Chemistry. 2nd edition.— Moscow, "Vysshaya shkola," 1968. (Russian)
13. Pasynskii, A.G.— Usp. Sovrem. Biol., No.3 (1957), 263.
14. Pasynskii, A.G.— Inform. Byull. "Radiobiologiya," No.6 (1964), 57.
15. Pasynskii, A.G.— Radiobiologiya 1 (1961), 3.
16. Pasynskii, A.G. and N.N.Demin.— Biokhimiya 25 (1961), 385.
17. Kuzin, A.M. Radiation Biochemistry.— Moscow, AN SSSR, 1962. (Russian)
18. Kuzin, A.M. The Structural and Metabolic Hypothesis in Radiobiology, p.112.— Moscow, "Nauka." 1970. (Russian)
19. Pasynskii, A.G. et al.— Radiobiologiya 6 (1964), 29.
20. Glik, M.C. and L.Warren.— Proc. Natn. Acad. Sci. U.S.A. 63 (1969), 563.
21. Tongur, A.M. and A.G.Pasynskii.— Radiobiologiya 7 (1967), 7.
22. Tongur, A.M. et al.— Radiobiologiya 10 (1970), 283.
23. Tongur, A.M. et al.— Radiobiologiya 10 (1970), 9.
24. Lewin, S.— J. Theoret. Biol. 23 (1969), 279.
25. Tongur, A.M. and A.G.Pasynskii.— Biofizika 5 (1960), 517.
26. Van der Eb, A.J. et al.— Biochim. biophys. Acta 182 (1969), 530.
27. Karpov, V.L. and B.I.Zverev.— In: Sbornik rabot po radiatsionnoi khimii, p.215. Moscow, AN SSSR, 1955.
28. Wallach, D.F.H. and A.Gordon.— Fedn. Proc. Fedn. Am. Soc. Exp. Biol. 27 (1968), 1263.
29. Zahler, P.H.— Experientia 25 (1969), 449.
30. Evans, R.J. et al.— Biochem. J. 7 (1968), 3095.
31. Deborin, G.A. et al.— Doklady AN SSSR 166 (1966), 231.
32. Triggle, D.J.— J. Theoret. Biol. 25 (1969), 499.
33. Day, C.E. and R.S.Levy.— J. Theoret. Biol. 22 (1969), 541.
34. Watkins, J.C.— Biochim. biophys. Acta 152 (1968), 293.
35. Tongur, A.M. et al.— Radiobiologiya. (In press)
36. Poltorak, O.M. and E.S.Vorob'eva.— Zh. Fiz. Khim. 40 (1966), 1665.
37. Chukhrai, E.S. and O.M.Poltorak.— Vestn. Mosk. Univ. 1 (1970), 10.
38. Emanuel', N.M.— Izv. AN SSSR, techn. sci. section 11 (1957), 1298.
39. Kudryashov, Yu.B.— In: "Pervichnye protsessy luchevogo porazheniya," p.90. Moscow, "Meditsina," 1966.
40. Tongur, A.M. et al.— Radiobiologiya. (In press)
41. Rothen, A.— J. Biol. Chem. 172 (1948), 841.
42. Hatchinson, F.— Archs. Biochem. Biophys. 41 (1952), 317.
43. Smith, C.L.— Archs. Biochem. Biophys. 43 (1954), 322.
44. Mazia, D. and G.Blumenthal.— J. Cell. Comp. Physiol. 35 (1950), 171.
45. Nadareishvili, K.Sh. and A.R.Egizarova.— Radiobiologiya 6 (1966), 503.
46. Kaplan, J.G. and M.J.Freser.— Biochim. biophys. Acta 9 (1952), 585.
47. Korgaonkar, K.S. and A.Desai.— Radiat. Res. 11 (1959), 625.
48. Desai, A.M. and K.S.Korgaonkar.— Radiat. Res. 21 (1964), 61.
49. Korgaonkar, K.S. et al.— Radiat. Res. 35 (1968), 213.
50. Egizarova, A.R. et al.— Radiobiologiya 6 (1966), 790.
51. Tkach, V.N. and V.V.Sidyakin.— Radiobiologiya 1 (1961), 641.
52. Kuzin, A.M. and N.B.Strazhevskaya.— Biofizika 1 (1956), 637.
53. Budnitskaya, E.V. et al.— Biokhimiya 23 (1958), 849.
54. Budnitskaya, E.V. et al.— Radiobiologiya 1 (1961), 37.
55. Echandi, R.J. and L.M.Massey.— Radiat. Res. 43 (1970), 372.
56. Hluchovsky, B. and V.Srb.— Biol. Zbl. 82 (1963), 73.
57. Srb, V. and B.Hluchovsky.— Expl. Cell. Res. 29 (1963), 261.
58. Srb, V.— Radiat. Res. 21 (1964), 308.
59. Srb, V.— Third Intern. Congress Radiat. Res. Cortina d'Ampezzo. Book of Abstr. Ref. 832, p.209, 26 June–2 July 1966.
60. Vasil'ev, I.M.— In: Deistvie ioniziruyushchikh izluchenii na rasteniya, p.26. Moscow, AN SSSR, 1962.
61. Brinkman, N.R. et al.— Int. J. Radiat. Biol. 3 (1961), 205.
62. Brinkman, N.R. et al.— Int. J. Radiat. Biol. 3 (1961), 509.

63. S c a i f e , J.F. and P.A l e x a n d e r.— Radiat. Res. 15 (1961), 658.
64. P e a r s e , A.G.E. Histochemistry, Theoretical and Applied. 2nd ed. Boston, Little Brown and Co. 1960.
65. G o r i z o n t o v a , M.P. and Z.A.T r e b e n o k.— Radiobiologiya 10 (1970), 690.
66. K i t a m u r o , T.— Shikoku Acta med. 21 (1965), 131.
67. H i l g e r t o v a , J. et al.— J. Nucl. Biol. Med. 10 (1966), 72.
68. T a r s h i s , M.A. et al.— Doklady AN SSSR 188 (1969), 700.
69. G o u t i e r , R.— Prog. Biophys. Chem. 11 (1961), 53.
70. G o u t i e r , R. and Z.M.B a c q.— In: R.M.Hochster and J.H.Quastel (Eds.), Metabolic inhibitors, Vol.2,
        p.631. New York, Academic Press, 1963.
71. D e D u v e , F.— In: L.Thomas, J. Uhr, and L.Grant (Eds.) Intern. Sympos. on Injury, Inflammation and
        Immunity, p.283. Baltimore, Williams and Wilkinson Co., 1964.
72. O k a d a , S. and L.D.P e a c h e y.— J. Biophys. Biochem. Cytol. 3 (1957), 239.
73. D e s a i , I.D. et al.— Biochim. biophys. Acta 86 (1964), 277.
74. W i l l s , E.D. and A.E.W i l k i n s o n.— Biochem. J. 99 (1966), 657.
75. H a r r i s , J.W.— Radiat. Res. 28 (1966), 766.
76. B r a n d e s , D. et al.— Cancer Res. 27 (1967), 731.
77. S o t t o c a s a , G.L. et al.— Radiat. Res. 24 (1965), 32.
78. W i l l s , E.D. and A.E.W i l k i n s o n.— Radiat. Res. 31 (1967), 732.
79. T s e n , C.C. and H.B.C o l l i e r.— Can. J. Biochem. Physiol. 38 (1960), 957.
80. P i e r u c c i , O. and W.R e g e l s o n.— Radiat. Res. 24 (1965), 619.
81. R a h m a n , Y.E.— Radiat. Res. 20 (1963), 741.
82. A c h e y , P.M. and E.C.P o l l a r d.— Radiat. Res. 31 (1967), 47.
83. P o l l a r d , E.C. and P.K.W e l l e r.— Radiat. Res. 35 (1968), 722.
84. B i l l e n , D.— Archs. Biochem. Biophys. 67 (1957), 333.
85. P o l l a r d , E.C. and P.K.W e l l e r. (Abstract).— Radiat. Res. 31 (1967), 617.
86. B h a u m i k , G. and S.B.Bh a t t a c h a r j e e.— Int. J. Radiat. Biol. 17 (1970), 101.
87. P o l l a r d , E.C. and C.V o g l e r — Radiat. Res. 15 (1961), 109.
88. G o l d f e d e r , A.— Trans. N.Y. Acad. Sci. 26 (1963), 215.
89. K u r n i c k , N. et al.— Radiat. Res. 11 (1959), 101.
90. G o u t i e r - P i r o t t e , M. and R.G o u t i e r.— Radiat. Res. 16 (1962), 728.
91. R o t h , J.S. and H.J.E i c h e l.— Radiat. Res. 9 (1958), 173.
92. S c a i f e , I.F.— Can. J. Biochem. Physiol. 41 (1963), 1469.
93. H a g e n , U.— Z. Naturf. 12b (1957), 546.
94. G o l d f e d e r , A. et al.— Int. J. Radiat. Biol. 12 (1967), 13.
95. F e o , F. et al.— Life Sci., Part 2, 9 (1970), 1235.
96. A r g u t i n s k a y a , S.V. and R.I.S a l g a n i k.— Radiobiologiya 5 (1965), 815.
97. C r e a s e y , W.A.— Biochem. J. 77 (1960), 5.
98. H a g e n , U. et al.— Biochim. biophys. Acta 74 (1963), 598.
99. L e D o u a r i n , G.H. and J.M.K i r r m a n n.— Third Intern. Congress Radiat. Res. Cortina d'Ampezzo,
        ref. 546, p.138, 1966.
100. T a k a m o r i , Y. et al.— Radiat. Res. 38 (1969), 551.
101. E r n s t , H.— Z. Naturf. 17B (1962), 300.
102. W h i t f i e l d , J.F. et al.— Int. J. Radiat. Biol. 9 (1965), 421.
103. C a m p a g n a r i , F. et al.— Expl. Cell Res. 42 (1966), 646.
104. S c a i f e , J.F. and B.H i l l.— Can. J. Biochem. Physiol. 40 (1962), 1025.
105. A l t e n b r u n n , H.J.— Radiobiol. Radiotherapie 5 (1964), 77.
106. S c h n e i d e r , H. et al.— Int. J. Radiat. Biol. 5 (1962), 183.
107. S m i r n o v a , T.V. et al.— Radiobiologiya 9 (1969), 751.
108. S a v i t s k i i , I.V.— In: "Mitokhondrii. Biokhimicheskie funktsii v sisteme kletochnykh organell," p.137.
        Sympos. Proceedings. Moscow, "Nauka." 1969.
109. N o y e s , P.P. and R.E.S m i t h.— Expl. Cell Res. 16 (1959), 15.
110. K a u f m a n , B.T. and N.O.K a p l a n.— Biokhim. biophys. Acta 39 (1960), 332.
111. A l t e n b r u n n , H.J. and E.Ko b b e r t.— Acta biol. med. germ. 9 (1962), 25.
112. M a x w e l l , E. and G.A s h w e l l.— Archs. Biochem. Biophys. 43 (1953), 389.
113. V a n B e k k u m , D.W.— In: "Ionizing Radiation and Cell Metabolism." Ciba Foundation Symposium.
        (G.E.Wolstenholme and C.M.O'Connor, eds.), p.77. Boston, Massachusetts, Little Brown, 1956.

114. Manoilov,S.E. and K.P.Khanson. — Vopr. Med. Khim. 4 (1964), 410.
115. Langendorff,H. and U.Hagen.— Experientia 11 (1955),485.
116. Scaife,J.F. and B.Hill.— Can. J. Biochem. Physiol. 41 (1963),1223.
117. Tarshis,M.A. et al.— Doklady AN SSSR 181 (1968),234.
118. Ivanitskaya,E.A. et al.— Biofizika 3 (1958),220.
119. Kuzin,A.M. and A.L.Shabadach.— In: Progress in Nuclear Energy, Ser.6, Biol. Sci.,Vol.2,p.364. London,Pergamon Press,1959.
120. Shabadash,A.L.— Radiobiologiya 1 (1961),212.
121. Vinetskii,Yu.P.— Radiobiologiya 2 (1962),370.
122. Burlakova,E.V. and O.R.Kol's.— Biofizika 3 (1958),711.
123. Polivoda,A.I. and A.A.Mikhailova.— Biofizika 5 (1960),612.
124. Nakhil'nitskaya,Z.N.— Radiobiologiya 5 (1965),25.
125. Kriger,Yu.A. and E.S.Elkhovskaya.— Biofizika 3 (1958),711.
126. Polivoda,B.I.— Radiobiologiya 9 (1969),488.
127. Polivoda,B.I.— Radiobiologiya 7 (1967),498.
128. Myers,D.K. and D.E.Slade.— Radiat. Res. 30 (1967),186.
129. Tribukait,B.— Nature 219 (1968),382.
130. Myers,D.K.— Int. J. Radiat. Biol. 19 (1971),293.
131. Whittam,R.— Biochem. J. 84 (1962),110.
132. Nakhil'nitskaya,Z.N.— Radiobiologiya 6 (1966),796.
133. Nakhil'nitskaya,Z.N.— In: "Fiziologiya i patologiya gistogematicheskikh bar'erov," p.298. Moscow, "Nauka." 1968.
134. Gerasimova,G.K. and Z.N.Nakhil'nitskaya.— Doklady AN SSSR 184 (1969),709.
135. Kövér,G. and E.Schoffeniels.— Int. J. Radiat. Biol. 9 (1965),461.
136. Shapiro,B. et al.— Radiat. Res. 27 (1966),139.
137. Shapiro,B. and G.Kollmann.— Radiat. Res. 34 (1968),335.
138. Myers,D.K. and R.W.Bide.— Radiat. Res. 27 (1966),250.
139. Kankura,T. et al.— Int. J. Radiat. Biol. 15 (1969),125.
140. Trincher,K.S.— Biofizika 4 (1959),78.
141. Trincher,K.S. and L.V.Orlova.— Radiobiologiya 5 (1965), 797.
142. Burdin,K.S. et al.— Radiobiologiya 5 (1965),922.
143. Woodbury,J.W.— Expl. Cell Res. 5 (1958),547.
144. Portela,A. et al.— Expl. Cell Res. 31 (1963),281.
145. Bergeder,H.D.— Biophysik,1 (1964),359.
146. Dawson,K.B. et al.— Radiat. Res. 37 (1969),83.
147. Ozerskii,M.I.— Radiobiologiya 9 (1969),189.
148. Gillet,G.— Int. J. Radiat. Biol. 8 (1964),533.
149. Wills,E.D.— Int. J. Radiat. Biol. 11 (1966),517.
150. Creasy,W.A.— Biochim. biophys. Acta 38 (1960),181.
151. Ting,T.P. and N.E.Zirkle.— J. Cell. Comp. Physiol. 16 (1940),197.
152. Sheppard,C.W. and G.E.Beyl.— J. Gen. Physiol. 34 (1951),691.
153. Cividalli,G.— Radiat. Res. 20 (1963),564.
154. Bresciani,F. et al.— Nature 196 (1962),186.
155. Bresciani,F. et al.— Radiat. Res. 22 (1964),463.
156. Sutherland,R.M. and A.Pihl.— Radiat. Res. 34 (1968),300.
157. Myers,D.K. et al.— Radiat. Res. 40 (1969),580.
158. Koefoed,V. Johnsen,and H.H.Ussing.— In: Mineral Metabolism (C.L.Comer and F.Bronner,eds.). Vol.1,p.169. New York, Academic Press,1960.
159. Myers,D.K. and L.Levy.— Nature 204 (1964),1324.
160. Myers,D.K. and M.L.Church.— Nature 213 (1967),636.
161. Sutherland,R.M. et al.— Int. J. Radiat. Biol. 12 (1967),551.
162. Pavlovskaya,T.E. et al.— Doklady AN SSSR 101 (1955), 723.
163. Pasynskii,A.G. et al.— Ibid.,p.317.
164. Jackson,K.L. and G.M.Christensen.— Radiat. Res. 27 (1966),434.
165. Gorizontova,M.P. et al.— Radiobiologiya 9 (1969),832.
166. Vasington,F.D. and J.V.Murphy.— J. Biol. Chem. 237 (1962),2670.

167. Rossi,C.S. and A.L.Lehninger.— J. Biol. Chem. 239 (1964),3971.
168. Chance,B.— J. Biol. Chem. 240 (1965),2729.
169. Rasmussen,H. and E.Ogata.— Biochemistry 5 (1966),733.
170. Azzi Azzone,G.F.— Biochim. biophys. Acta 113 (1966),445.
171. Chappel,J.B. and G.D.Greville.— Fedn. Proc. Fedn. Am. Soc. Exp. Biol. 22 (1963),526.
172. Merkulov,A.S.— Thesis. Pushchino-na-Oke,1969.
173. Kalacheva,V.Ya. and T.E.Pavlovskaya.— In press.
174. Kalacheva,V.Ya. and N.M.Sisakyan.— Biokhimiya 30 (1965),858.
175. Parsons,D. et al.— Ann. N.Y. Acad. Sci. 137 (1966),643.
176. Bachmann,E. et al.— Archs. Biochem. Biophys. 115 (1966),153.
177. Bachmann,E. et al.— Archs. Biochem. Biophys. 121 (1967),73.
178. Sottocasa,G.L. and G.Sandri.— Ital. J. Biochem. 27 (1968),17.
179. Frank,G.M.— Biofizika 15 (1970),298.
180. Korn,E.D.— In: Ann. Rev. Biochem. 38 (1969),263.
181. Molotkovskii,Yu.G.— Fiziologiya Rastenii 17 (1970),1249.
182. Potter,R.L. and F.H.Bethell.— Fedn. Proc. Fedn. Am. Soc. Exp. Biol. 11 (1952),270.
183. Zaslvaskii,Yu.A. and V.N.Shchipakin.— In: "Mitokhondrii. Biokhimicheskie funktsii v sisteme
       kletochnykh organell," p.94. Sympos. Mater. Moscow, "Nauka." 1969.
184. Yang Fu-Yu et al. Scientia Sinica,12 (1963),1595.
185. Yang Fu-yu et al. Acta Biol. Exper. Sinica,9 (1964),261.
186. Nitz Litzow,D. and G.Bührer.— Strahlentherapie 113 (1960),201.
187. Khanson,K.P.— Thesis. Leningrad,1964.
188. Khanson,K.P.— Radiobiologiya 5 (1965),44.
189. Urakami,H.— Okayama-Igakkai-Zasshi 76 (1965),13.
190. Yost,M.T. et al.— Radiat. Res. 32 (1967),187.
191. Meisel', M.N.— In: "Deistvie oblucheniya na organizm," p.78. Moscow, AN SSSR, 1955.
192. Sisakyan,N.M. and V.Ya.Kalacheva.— Biokhimiya 26 (1961),277.
193. Thomson,G.— Radiat. Res. 21 (1964),46.
194. Voskoboinikov,G.V.— Izv. AN SSSR,biol. ser. 6 (1969),832.
195. Kondrashova,M.N. Mitochondria. Biochemistry and Morphology,p.137.— Moscow, "Nauka." 1967.
       (Russian)
196. Kondrashova,M.N.— Biofizika 15 (1970),312.
197. Romantsev,E.F. et al.— Radiobiologiya 9 (1969),213.
198. Romani,R.J. and I.K.Yu.— Archs. Biochem. Biophys. 117 (1966),638.
199. Joshi,V.G. and V.K.Gaur.— Int. J. Radiat. Biol. 18 (1970),173.
200. Romani,R. et al.— Radiat. Bot. 7 (1967),41.
201. Miller,L.A. et al.— Radiat. Bot. 7 (1967),47.
202. Romani,R.J. and J.B.Biale.— Pl. Physiol. 32 (1957),662.
203. Vines,H.M.— Pl. Physiol. 39 (1964),225.
204. Khanson,K.P.— Vopr. Med. Khim. 12 (1966),256.
205. Benjamin,T.L. and H.T.Yost.— Radiat. Res. 12 (1960),613.
206. Pozar,B.L.— In: Symposium on Genetics and Wheat Breeding. J.Rajki (ed.),p.237. Martonvasar,
       Hungary, Agric. Res. Inst. Hungarian Acad. Sci.,1963.
207. Zicha,B. et al.— Experientia 22 (1966),712.
208. Kuzin, A.M. and V.A.Kopylov.— Radiobiologiya 2 (1962),681.
209. Lehninger,A.L.— In: "Molecular Organization and Biological Function." J. McAllen (ed.),p.107.
       New York,1967.
210. Goldfeder,A. et al.— J. Cell. Biol. 23 (1964),118A.
211. Manteifel',V.M. and M.N.Meisel'.— Radiobiologiya 2 (1962),101.
212. Manteifel',V.M. and M.N.Meisel'.— Izv. AN SSSR,biol. ser. 6 (1965),884.
213. Scherer,E. and V.Vogell.— Strahlentherapie 106 (1958),202.
214. Braun,H.— Strahlentherapie 117 (1962),134.
215. Ferents,A.I.— Radiobiologiya 10 (1970),658.
216. Braun,H.— Naturwissenschaften 45 (1958),18.
217. Siebert,G.— In: "The Cell Nucleus, Metabolism and Radiosensitivity," p.265. London, Taylor, 1966.
218. Betel,J. and H.M.Klouwen.— In:"Cell Nucleus. Metabolism and Radiosensitivity," p.281. London,
       Taylor,1966.

219. McEwen, B.C. et al.– J. Biol. Chem. 238 (1963), 758.
220. Betel, I. and H.M.Klouwen.– Biochem. biophys. Acta 131 (1967), 453.
221. Conover, T.E.– 7th Intern. Congr. Biochem. Abstr. 889, 1967.
222. Allfrey, V.G. et al.– Nature 176 (1955), 1042.
223. Osawa, S. et al.– J. Gen. Physiol. 40 (1957), 491.
224. Creasey, W.A. and L.A.Stocken.– Biochem. J. 69 (1958), 17.
225. Creasey, W.A. and L.A.Stocken.– Biochem. J. 72 (1959), 519.
226. Ord, M.G. and L.A.Stocken.– Biochem. J. 84 (1962), 593.
227. Klouwen, H.M. et al.– Int. J. Radiat. Biol. 6 (1963), 441.
228. Klouwen, H.M. et al.– In: "Cell Nucleus. Metabolism and Radiosensitivity," p.295. London, Taylor, 1966.
229. Zimmerman, D.H. and H.L.Cromroy.– Life Sci. 6 (1967), 621.
230. Klouwen, H.M.– In: "Cellular Radiation Biology," p.142. Baltimore, Williams and Wilkins Co., 1965.
231. Klouwen, H.M. et al.– Nature 209 (1966), 1149.
232. Betel, I.– Int. J. Radiat. Biol. 12 (1967), 459.
233. Araki, K. et al.– Int. J. Radiat. Biol. 17 (1970), 375.
234. Kuzin, A.M. and M.A.Tarshis.– Radiobiologiya 9 (1969), 755.
235. Kuzin, A.M. and M.A.Tarshis.– Radiobiologiya 10 (1970), 116.
236. Tongur, A.M. et al.– Second All-Union Biochem. Congress, Tashkent, 1969, Sec. 17, p.44. FAN UzbSSR, 1969. (Russian)
237. Delektorskaya, L.N. and K.A.Perevoshchikova.– Biokhimiya 34 (1969), 199.
238. Ohnishi, T. et al.– J. Biochem. 50 (1964), 6.
239. Klein, R.L. and B.A.Afzelius.– Nature 212 (1966), 609.
240. Yasuzumi, G. and I.Tsubo.– Expl. Cell Res. 43 (1966), 281.
241. Yasuzumi, G. et al.– Expl. Cell Res. 45 (1967), 261.
242. Yasuzumi, G. et al.– J. Ultrastruct. Res. 23 (1968), 321.
243. Braun, H.– Strahlentherapie 122 (1963), 248.
244. Karupu, V.Ya. et al.– Proceedings of the Third All-Union Sympos. on the Structure and Function of the Cell Nucleus, p.50, Kiev, "Naukova Dumka," 1970. (Russian)
245. Nadareishvili, K.Sh. et al.– Radiobiologiya 8 (1968), 396.
246. Nadareishvili, K.Sh. et al.– Radiobiologiya 10 (1970), 364.
247. Tarusov, B.N. Principles of the Biological Effect of Radioactive Radiation.– Moscow, Medgiz, 1954. (Russian)
248. Zhuravlev, A.I. In: "Proceedings of the Scientific Conference on the Pathogenesis, Therapy and Propylaxis of Radiation Sickness," p.33. Leningrad, 1957. (Russian)
249. Kudryashov, Yu.B.– Thesis. Moscow, 1966.
250. Kakushkina, M.L. et al.– Vopr. Med. Khim. 12 (1966), 147.
251. Budnitskaya, E.V. and I.G.Borisova.– In: "Rol' perekisei i kisloroda v nachal'nykh stadiyakh radiobiologicheskogo effekta," p.85. Moscow, 1960.
252. Budnitskaya, E.V. and I.G.Borisova.– Biokhimiya 26 (1961), 142.
253. Borisova, I.G.– Thesis. Moscow, 1966.
254. Budnitskaya, E.V. et al.– Doklady AN SSSR 120 (1958), 140.
255. Wills, E.D. and A.E.Wilkinson.– Int. J. Radiat. Biol. 17 (1970), 229.
256. Robinson, M.G. et al.– Biochim. biophys. Acta 124 (1966), 181.
257. Scott, D.B.M. et al.– Biochem. J. 93 (1964), 1c.
258. Hagen, U.– Naturwissenschaften 47 (1960), 601.
259. Hagen, U.– Strahlentherapie 117 (1962), 119.
260. Hagen, U.– Nature 187 (1960), 1123.
261. Bauer, R.D. et al.– Radiat. Res. 20 (1963), 24.
262. Bauer, R.D. et al.– Int. J. Radiat. Biol. 16 (1969), 575.
263. Goutier, R. et al.– Int. J. Radiat. Biol. 8 (1964), 51.
264. Whitfield, J.F. et al.– Expl. Cell Res. 36 (1964), 341.
265. Perris, A.D. and J.F.Whitfield.– Int. J. Radiat. Biol. 11 (1966), 399.
266. Perris, A.D. et al.– Expl. Cell Res. 45 (1967), 48.
267. Kuzin, A.M. et al.– Radiobiologiya 10 (1970), 103.
268. Ernst, H.– Z. Naturf. 16B (1961), 329.
269. Ernst, H.– Z. Naturwiss. 48 (1961), 575.

270. Ernst, H.— Z. Naturwiss. 50 (1963), 333.
271. Swingle, K.F. and L.J.Cole.— Radiat. Res. 39 (1969), 483.
272. Korol', B.A. et al.— Radiobiologiya 11 (1971), 180.
273. Ryser, H.J.P. and R.Hancock.— Science 150 (1965), 501.
274. Hancock, R. and H.J.P.Ryser.— Nature 213 (1967), 701.
275. Tarshis, M.A. et al.— Tsitologiya 12 (1970), 794.
276. Schwartz, A. and C.L.Johnson.— Life Sci. 4 (1965), 1555.
277. Schwartz, A. et al.— J. Biol. Chem. 241 (1966), 4505.
278. Johnson, C.L. et al.— J. Biol. Chem. 241 (1966), 4513.
279. Johnson, C.L. et al.— Biochemistry 7 (1967), 1121.
280. Harris, E.J. et al.— Ibid., p.1360.
281. Schwartz, A.— J. Biol. Chem. 240 (1965), 939.
282. Dorgii, I.E. et al.— Radiobiologiya 10 (1970), 836.
283. Nemchinskaya, I.E. et al.— Radiobiologiya 30 (1965), 33.
284. Braun, A.D. et al.— Tsitologiya 7 (1965), 494.
285. Canelina, L.Sh. et al.— Tsitologiya 8 (1966), 526.
286. Schwartz, A.— J. Biol. Chem. 240 (1965), 944.
287. Pisareva, L.N. et al.— Tsitologiya 12 (1970), 1405.
288. Schwartz, A. and A.H.Laseter. — Life Sci. 4 (1965), 145.
289. Schwartz, A.— Biochim. biophys. Acta 100 (1965), 202.
290. Person, P. and A.Fine.— Archs. Biochem. Biophys. 94 (1961), 392.
291. Berezney, R. et al.— Biochem. Biophys. Res. Commun. 38 (1970), 93.
292. Braun, A.D.— Proceedings of the Second All-Union Biochem. Congress, p.219. Tashkent, FAN, 1969.

SUBJECT INDEX